HYBRID IMAGE PROCESSING METHODS FOR MEDICAL IMAGE EXAMINATION

Intelligent Signal Processing and Data Analysis
Series Editor: Nilanjan Dey

Intelligent signal processing (ISP) methods are progressively swapping the conventional analog signal processing techniques in several domains, such as speech analysis and processing, biomedical signal analysis radar and sonar signal processing, telecommunications, and geophysical signal processing. The main focus of this book series is to find out the new trends and techniques in the intelligent signal processing and data analysis leading to scientific breakthroughs in applied applications. Artificial fuzzy logic, deep learning, optimization algorithms, and neural networks are the main themes.

Bio-Inspired Algorithms in PID Controller Optimization
Jagatheesan Kallannan, Anand Baskaran, Nilanjan Dey, Amira S. Ashour

A Beginner's Guide to Image Preprocessing Techniques
Jyotismita Chaki, Nilanjan Dey

Digital Image Watermarking: Theoretical and Computational Advances
Surekha Borra, Rohit Thanki, Nilanjan Dey

A Beginner's Guide to Image Shape Feature Extraction Techniques
Jyotismita Chaki, Nilanjan Dey

Coefficient of Variation and Machine Learning Applications
K. Hima Bindu, Raghava Morusupalli, Nilanjan Dey, C. Raghavendra Rao

Data Analytics for Coronavirus Disease (COVID-19) Outbreak
Gitanjali Rahul Shinde, Asmita Balasaheb Kalamkar, Parikshit Narendra Mahalle, Nilanjan Dey

A Beginner's Guide to Multi-Level Image Thresholding
Venkatesan Rajinikanth, Nadaradjane Sri Madhava Raja, Nilanjan Dey

Hybrid Image Processing Methods for Medical Image Examination
Venkatesan Rajinikanth, E. Priya, Hong Lin, Fuhua (Oscar) Lin

For more information about this series, please visit: https://www.routledge.com/Intelligent-Signal-Processing-and-Data-Analysis/book-series/INSPDA

HYBRID IMAGE PROCESSING METHODS FOR MEDICAL IMAGE EXAMINATION

Venkatesan Rajinikanth, E. Priya, Hong Lin, and Fuhua Lin

CRC Press
Taylor & Francis Group
Boca Raton London New York

CRC Press is an imprint of the
Taylor & Francis Group, an **informa** business

First edition published 2021
by CRC Press
6000 Broken Sound Parkway NW, Suite 300, Boca Raton, FL 33487-2742

and by CRC Press
2 Park Square, Milton Park, Abingdon, Oxon, OX14 4RN

First edition published by CRC Press 2021

CRC Press is an imprint of Taylor & Francis Group, LLC

ISBN: 9780367534967 (hbk)
ISBN: 9781003082224 (ebk)

Typeset in Times New Roman
by MPS Limited, Dehradun

Contents

Preface

Recent improvements in science and technology helped to develop a considerable number of medical facilities to detect diseases in their premature phase. These improvements also provided an appropriate treatment process to control and cure diseases. Most internal organ diseases are normally assessed with the help of medical images recorded with varied modalities; hence, an appropriate image processing system is essential to detect the disease and its severity from available medical images.

Improving the information obtainable in untreated digital illustration is generally performed with a selected image improvement practice. The image improvement techniques play a vital role in analyzing a variety of imaging modalities, such as fundus retinal images, blood slides, dermoscopic images, ultrasounds, mammograms, thermal imaging, CT scan slices, X-rays, and MRI slices. Recently, image-assisted disease detection and treatment planning improved medical industries, and a considerable number of image examination schemes are proposed and implemented by researchers. In medical image analysis, appropriate pre-processing and post-processing techniques are implemented to extort and appraise the disease-infected segment from the digital image. Further, the overall accuracy of this disease detection system depends on the pre-processing process. Hence, in this book, the commonly used image pre-processing technique called the 'multi-thresholding process' is discussed with appropriate examples. The implementation of the traditional and heuristic algorithm-based disease detection system is also discussed with appropriate examples. A detailed study with the hybrid image processing methods and the deep-learning based automated disease classification is also presented. Finally, the implementation of the deep-learning system is demonstrated using the lung CT scan slices of the COVID-19 dataset. In this work, the proposed work is experimentally demonstrated using MATLAB® and Python software.

The book is organized as follows:

Chapter 1 presents an overview of the disease screening process followed in hospitals to analyze and confirm disease in various internal and external human organs. This section also presents the procedures followed to screen COVID-19 infected patients during disease diagnosis and confirmation. Further, this section presents a detailed discussion regarding the image recording procedures followed in hospitals to analyze the disease.

Chapter 2 demonstrates the need for image enhancement procedures with appropriate experimental results obtained with MATLAB. The need for improvement is outlined briefly by well-known methods such as artifact removal, filtering, contrast enhancement, edge detection, thresholding, and smoothing. The recent advancement in image enhancements, such as the hybrid image assessment technique, is also presented with experimental results.

Chapter 3 discusses details on the choice of suitable image examination procedures that are demonstrated using appropriate results attained using MATLAB software. Further, this chapter presents the details regarding particle swarm optimization, bacterial foraging, firefly, bat, cuckoo, social group optimization, teaching-learning, and the jaya algorithm and their role during the image thresholding process.

Chapter 4 presents an overview of the traditional and the CNN-based segmentation procedures. This section also presents the experimental outcome of the proposed technique on greyscale and RGB images with and without noise. The performance evaluation of the proposed segmentation is also demonstrated using appropriate examples.

Chapter 5 demonstrates the implementation of the hybrid image processing technique implemented to examine brain tumors using brain MRI slices. This section presents a detailed demonstration of the various traditional segmentation procedures considered to extract the tumor section from the MRI slice.

Chapter 6 presents an overview of deep-learning architectures such as AlexNet, VGG-16, and VGG-19 and their application in medical image classification tasks. This section further discusses the transfer-learning technique and the essential modification to be implemented to enhance the classification accuracy. A detailed lung CT scan slice classification with the VGG architecture is demonstrated using the public image dataset collected from COVID-19 patients. This section presents the experimental result attained using MATLAB and Python software.

Chapter 7 concludes the presented work in the previous chapters and discusses the future scope.

Dr. Venkatesan Rajinikanth,
St. Joseph's College of Engineering
Dr. E. Priya,
Sri Sairam Engineering College
Dr. Hong Lin,
University of Houston-Downtown
Dr. Fuhua (Oscar) Lin,
Athabasca University

Authors

Dr. Venkatesan Rajinikanth is a professor in the Department of Electronics and Instrumentation Engineering at St. Joseph's College of Engineering, Chennai, Tamilnadu, India. Recently, he edited books titled *Advances in Artificial Intelligence Systems* (Nova Science Publishers, USA) and *Applications of Bat Algorithm and its Variants* (Springer, Singapore). He is the Associate Editor of the *International Journal of Rough Sets and Data Analysis* (IGI Global, US, DBLP, ACM dl) and has edited special issues in journals: *Current Signal Transduction Therapy, Current Medical Imaging Reviews*, and *Open Neuroimaging Journal*. His main research interests include medical imaging, machine learning, and computer-aided diagnosis, as well as data mining.

Dr. E. Priya is a professor in the Department of Electronics and Communication Engineering, Sri Sairam Engineering College, Chennai, Tamilnadu, India. She is currently guiding students in the areas of biomechanical modeling and image & signal processing. Her research interests include biomedical imaging, image processing, signal processing, and the application of artificial intelligence and machine learning techniques. A recipient of the DST-PURSE fellowship, she has published several articles in international journals and conference proceedings, as well as book chapters, in the areas of medical imaging and infectious diseases. She recently edited the book *Signal and Image Processing Techniques for the Development of Intelligent Healthcare Systems* (Springer Nature). She also serves on the editorial review board of the *International Journal of Information Security and Privacy* (IJISP), IGI Global.

Dr. Hong Lin received his PhD in Computer Science in 1997 from the University of Science and Technology of China. Before he joined the University of Houston-Downtown (UHD) in 2001, he was a postdoctoral research associate at Purdue University; an assistant research officer at the National Research Council, Canada; and a software engineer at Nokia, Inc. Dr. Lin is currently a professor in Computer Science with UHD. His research interests include cognitive intelligence, human-centered computing, parallel/distributed computing, and big data analytics. He is the supervisor of the Grid Computing Lab at UHD. He is also a senior member of the Association for Computing Machinery (ACM).

 **Dr. Fuhua Lin** is a professor in the School of Computing and Information Systems, Faculty of Science and Technology of Athabasca University, Canada. His current research interest is applying AI technology to enhance online learning environments, e.g., incorporating adaptivity into virtual worlds for Cyber-learning to build an engaging, personalized, and interactive online learning environment. He has more than 100 publications, including edited books, journal papers, book chapters, conference papers, and reviews. He is the editor of the book *Designing Distributed Learning Environments with Intelligent Software Agents*, published by Information Science Publishing. Dr. Lin was the Editor-In-Chief of the *International Journal of Distance Education Technologies*. He is also a senior member of the Association for Computing Machinery (ACM) and IEEE.

1 Introduction

Recent developments in the science, technology, and medical domains have helped people live better lives. This improved lifestyle is also due to access to a wide variety of facilities, including state-of-the-art medical facilities. Further, a considerable number of vaccinations and preventive medication has helped reduce infectious and communicable diseases. Advanced treatment facilities available in multi-speciality hospitals have also helped in detecting and curing diseases in their early stages. Moreover, scheduled health checkups recommended by doctors also help to identify and cure a number of acute and deadly diseases.

Although considerable measures have been taken to prevent and cure human diseases, the incidence rate of new kinds of infections and communicable/non-communicable diseases is rapidly rising, irrespective of age, race, and gender. To support the early diagnosis and treatment implementation for diseases, a number of diagnostic procedures are proposed and implemented in various disease diagnostic centres and clinics. The medical images recorded using a chosen modality gives required insight for the disease to be identified. Based on this insight, the doctor can plan what treatment is to be implemented [1,2].

Medical imaging can be recorded with a variety of procedures ranging from camera-assisted techniques to radiation-based methods. The choice of methodology depends upon the organ to be examined and the expertise of the doctor. The recorded image can be assessed by the doctor, assisted by a computerised algorithm. Modern image recording facilities support information in digital form, known as digital imaging, thus computer-assisted diagnosis is widely adopted. Aside from this, digital images can be easily stored, retrieved, and processed using a number of techniques [3].

This chapter presents an overview of disease screening procedures for a chosen disease. The invasive and non-invasive image recording procedures existing present in literature and the assessment of the images are discussed briefly along with appropriate examples.

At the present, although a considerable number of medical facilities are available, disease occurrence rates among humans are gradually rising due to various unavoidable causes. These illness in humans can be classified into several categories, such as: (i) infectious disease, (ii) deficiency disease, (iii) hereditary disease, and (iv) physiological disease. The above mentioned diseases can further be grouped as communicable and non-communicable.

Diseases occurring externally in the human body are quite easy to detect and treat when compared with the diseases found in internal organs. A pre-screening procedure is recommended as essential to detect the disease in its premature phase. If the disease and its severity are identified in its premature phase, a treatment procedure could be recommended and implemented to control and cure the disease. This could be could imply less effort compared to a disease diagnosed in its delayed phase.

Diseases found in external body organs, such as the eye and skin, could be examined with a personal checkup by an experienced doctor along with an image supported detection system for further examination. Meanwhile, diseases of internal organs such as the brain, lungs, heart, breast, the digestive system, and blood are normally diagnosed using a chosen imaging method associated with a prescribed imaging modality. The imaging procedure followed for the internal organ needs complete monitoring and should be examined in a controlled environment with a prescribed clinical protocol. After registering the image of the organ, the disease can be diagnosed using a computerized disease examination procedure or a personal check by a clinical expert.

In most cases, semi-automated/automated disease detection procedures are implemented to speed up the diagnostic process. The report prepared with these techniques are used as supporting evidence regarding the patient, which will help the doctor during decision making and treatment planning process. Further, the availability of the computing facility helps to develop a large number of computer-aided detection procedures, which considerably reduce the burden on doctors during conventional disease detection and also during mass screening processes.

This book aims to discuss the details of the Artificial-Intelligence (AI) based disease detection procedures mainly developed by Machine-Learning (ML) and Deep-Learning (DL) techniques. Further, this book also presents the details of Hybrid Image Processing (HIP) methods implemented to enhance the detection accuracy for a class of clinical images.

1.1 INTRODUCTION TO DISEASE SCREENING

Diseases are medical emergencies where the unrecognised and untreated will cause various problems, including death. These diseases are classified either as communicable or non-communicable depending on the occurrence rate and its nature. Diseases of the human body can be diagnosed in a variety of procedures, and a visual check is preferred due to its accessibility. The disease detection procedure in external organs is easy while also helping identify the severity. In some moderate/ acute cases, along with a personal check, a suitable signal/image-based disease evaluation is also recommended by the doctor to verify and confirm the disease.

Diseases in vital internal organs are more when compared with external organs, hence more care needs to be taken during diagnosis. Most of the diseases in internal organs such as the heart, lungs, brain, kidney, respiratory tract, stomach, and blood are normally diagnosed using carefully chosen bio-signalling/bio-imaging procedures. The bio-imaging-based assessment helps attain more insight regarding the organs to be examined compared to the signal-based techniques. Hence, medical imaging-assisted disease diagnosis has emerged as a common and recommended technique. In this method, an image modality is considered to record and examine the internal organ.

Developments in medical imaging techniques helped to achieve a two-dimensional (2D) or the three-dimensional (3D) picture of the organ in greyscale or RGB form. These images would help the physician to get an insight regarding the disease in the body and also helps to track the progression of the disease with respect to time. Identification of the disease in its premature phase is very essential to plan for the appropriate treatment. Treatment of the premature phase disease is easy compare to other stages and hence a number of scheduled screening procedures could be planned and conducted at an early stage. The scheduled body screening will help to identify number of diseases in its premature phase, even though the symptoms are absent.

As discussed earlier, the disease in humans can be commonly classified as (i) communicable and (ii) non-communicable diseases. Each disease will have its own symptom; and the patient will immediately approach the physician when he/she experiences a disease symptom. The doctor will examine the patient with the prescribed protocol existing to identify the disease based on the symptom as well as the difficulty experienced by the patient.

The doctor will suggest a range of preliminary diagnostic procedures to confirm the disease and assess the severity level. The procedures executed to test the patient for confirmation of the disease is technically known as Disease-Screening (DS) process and it varies according to the disease to be detected. The procedures commonly employed in DS involves (i) personal check by an expert, (ii) clinical test ranging from sample collection and testing of the bio-signal/bio-image-based methods, (iii) intermediate level detection based on the bio-signals and images collected from the patients, and (iv) verification of report by the doctor for authentication of disease.

The overview of the clinical level diagnosis of the disease in human is presented in this section with appropriate block diagrams.

Figure 1.1 illustrates the initial-level verification and recommendation by the doctor a patient approaches. This protocol includes common demographic parameters such as gender, age and weight, followed by previous history of disease, the number of days the patient has been afflicted by the symptoms, heart rate, temperature, and other recommended parameters to identify the cause and nature of the disease. If the disease is in any internal organ then along with personal verification, the doctor will also suggest a clinical checkup to collect more information about the disease. The doctor can then plan for appropriate treatment to control the disease and take necessary steps for curing after careful inspection of the clinical report.

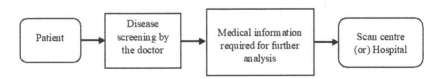

FIGURE 1.1 Initial Level Verification and Recommendation by the Doctor.

Figure 1.2 depicts various phases involved in the recording of diseases in vital internal organs using the appropriate imaging modality. Diseases in the internal organs are initially evaluated by a doctor and, for further assessment, the image of the organ is recorded with a chosen modality. To maximize quality and maintain confidentiality, the image recording procedure is performed in a controlled environment where all the prescribed protocols are followed during recording. This also helps avoid ethical and legal issues. After recording the image, an initial-level assessment is performed in the scan/imaging centre. The recorded image, along with the prediction during the imaging, is then sent to the doctor for further evaluation. The doctor examines the image and report and, based on the observation, a conclusion regarding the orientation, cause and harshness of the disease can be computed.

The general image processing system presented in Figure 1.3 assesses the medical images recorded by the chosen modality. The raw image recorded from the patient using standard acquisition protocol undergoes a variety of pre-processing procedures based on what is needed. Some of the commonly used image pre-processing techniques include (i) image orientation adjustment, (ii) resizing, and (iii) filtering. After pre-processing, a post-processing technique may then be employed to extract the essential section or the necessary features from the image. This procedure assists the doctor in diagnosing the disease using a dedicated computerised software.

Due to its clinical significance, several semi-automated and automated disease detection systems are implemented by the researchers to diagnose diseases. These disease diagnostic tools aid the doctor in making prompt decisions during the treatment process.

The upcoming sections of the book present the various imaging modalities considered to record images (gray/RGB scale) related to the diseases. The images presented are collected from well known image datasets existing in literature. Every image, as well as existing evaluation procedures, is also discussed in these subsections.

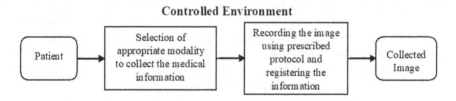

FIGURE 1.2   Collection of Disease Information Using a Prescribed Protocol.

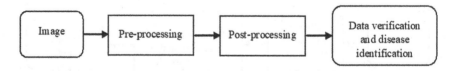

FIGURE 1.3   Image Assessment Technique to Convert the Raw Information into Understandable Information.

1.1.1 SCREENING OF BLOOD SAMPLE

Blood sample screening is a commonly recommended procedure by doctors during routine health examinations. It is handled as an initial-level screening for infections and communicable diseases. Blood testing is a common lab analysis technique widely performed to measure the blood cell count and the infection in the cells. An essential amount of blood is collected from the patient during this process as per the prescribed protocol. This is then used to evaluate the Complete Blood Count (CBC), blood chemistry, and blood enzymes. This section presents the image-assisted assessment of White Blood Cell (WBC) detection for Malarial and *Trypanosoma cruzi* (*T. cruzi*) infection.

Once sufficient volume of the blood is collected from the patent using the prescribed protocol, the collected blood is converted into blood film/peripheral blood smear (thin/thick) using the glass microscope slide. These are then marked in such a way as to observe different blood cells. The marking/staining agents are used to enhance the visibility of the information to be collected from the peripheral blood smear.

- **White Blood Cell**

The WBC, also known as leukocytes, is a major component of the immune system. It protects the body from infectious microorganisms and foreign intruders. There are certain categories of WBC that are generated from the hematopoietic stem cells of bone marrow. The WBCs are found throughout the body due to its lymphatic structure [4,5]. During the disease screening process, the WBC/leukocyte count plays a vital role. This count is normally measured using the universal unit "cells per microliter (cells/mcL)" where the normal count ranges from 4,000 to 11,000 cells/mcL. A decrease in the value indicates a lower immunity level, and an increase presents the possibility of infection. As per the literature, there are five WBC categories present in the human body, each having its own function and texture pattern. The clinical-level recording of the WBC is achieved using digital microscopes. The recorded microscopic image is then assessed using a suitable WBC detection system. Various WBC cells prepared from thin blood smear viewed under a digital microscope are presented in Figure 1.4. These sample images were collected from the benchmark database known as Leukocyte Images for Segmentation and Classification (LISC) [6,7]. The thin blood smear images of WBCs in Figure 1.4 (a) to (e) are the Basophil, Eosinophil, Lymphocyte, Monocyte and Neutrophil. Figure 1.4 (f) presents the image of a mixed WBC recorded using a thin blood smear.

During the image acquisition process, blood samples collected from the peripheral blood of 8 healthy and 400 infected patients are smeared on the slides and viewed under a microscope. These slides are smeared with Gismo-Right technique and viewed with a light microscope (Microscope-Axioskope 40) using an achromatic lens with a magnification factor of 100. The view fields are then captured by a digital camera (Sony Model No. SSCDC50AP) and are saved as .bmp files with a pixel resolution of $720 \times 576 \times 3$. The earlier research work carried out using the LISC data can be found in [7].

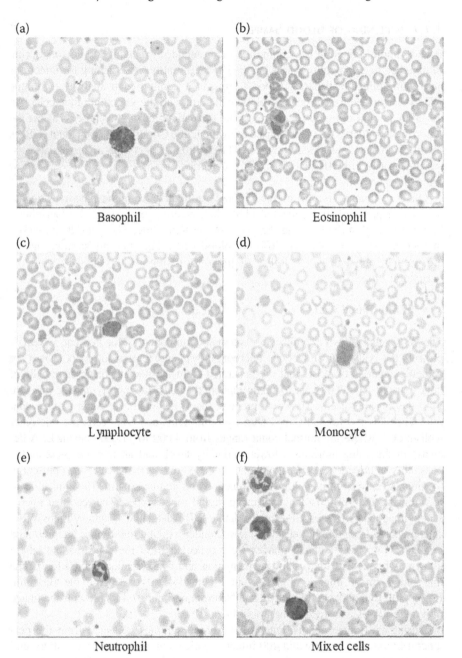

(a)

Basophil

(b)

Eosinophil

(c)

Lymphocyte

(d)

Monocyte

(e)

Neutrophil

(f)

Mixed cells

FIGURE 1.4 Sample Test Images of White Blood Cells (Leucocytes).

Advantages: The assessment of leukocyte type and its count in cells/mcL helps confirm infection in the human body and is one of the commonly accepted pre-screening processes for a variety of diseases, including cancer and COVID-19.

Limitations: This procedure requires the collection of a prescribed volume of blood from infected patients. Further, blood collection and preparation of the microscopic slide and staining requires care to avoid spread of the disease.

- **Malaria**

In most countries, mosquito-borne infectious diseases are common and can cause moderate to severe symptoms among humans. Malaria is one of the diseases caused by the single-celled microorganism Plasmodium. Even though a variety of plasmodium species exist, only five groups cause malaria. If malaria is identified in its early phase, a possible treatment procedure can be implemented to help the patient recover. Malaria infection develops mostly either from the liver or Red Blood Cell (RBC). A mosquito bite injects the plasmodium parasite into the body which later will reach and settle in the RBC. Within the RBC, the plasmodium multiplies further and causes severe illness to the patient [8,9].

It is essential to identify the severity of RBC infection and the type of Plasmodium which caused the illness to implement a possible treatment process. To achieve this, blood sample collection and microscopic examination with the blood film/peripheral blood smear (thin/thick) are normally performed in a clinical laboratory. After the identification of the infection level and the type of the parasite, the doctor can treat and prescribe the suitable drug for recovery [10].

Figure 1.5 presents the thin blood smear microscopic images of common Plasmodium parasites that infect humans, such as *P. falciparum*, *P. malariae*, *P. vivax*, *P. ovale* and *P. knowlesi*. Literature confirms that malaria due to *P. falciparum* is acute to humans compared to malarial infection by other species. Further, the hypnozoites developed by *P. vivax* remain in the body for several years. Malaria is one of the deadliest diseases in Africa, Asia and Latin America. Earlier research related to malaria and the image-assisted diagnostic techniques can be referred to in [8,9]. Figure 1.5 (a) to (e) presents parasites such as *P. falciparum*, *P. malariae*, *P. vivax*, *P. ovale,* and *P. knowlesi* acquired from a thin blood smear.

Advantages: Malaria parasite and its infection level could be accurately detected and identified using microscopes. Further, this technique is a traditional and accepted clinical practice to detect blood infections.

Limitations: This examination requires the gathering of an approved volume of blood from the infected person. Further, the preparation of the microscopic slide and staining, as well as the disposal of the collected blood, requires utmost care.

- **Chagas disease**

Chagas disease, also known as American trypanosomiasis, is a vector-borne disease spread by the Triatominae bug. The infected bug normally carries the parasite called *Trypanosoma cruzi* (*T. cruzi*). When the bug bites, it injects the parasite into the blood stream where it can grow and multiply. This disease severely affects the blood tract, including the heart If left untreated, the disease can cause various illnesses [11].

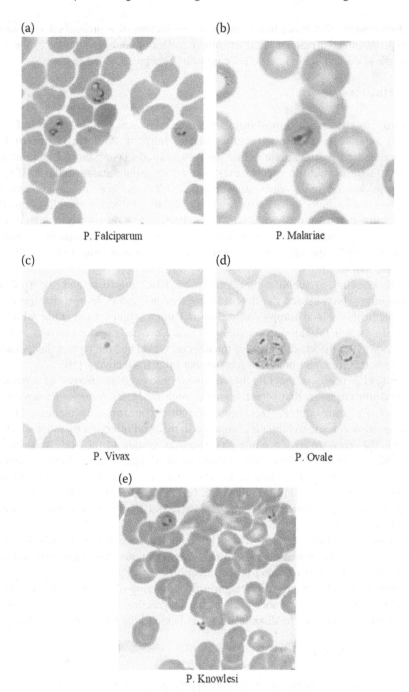

FIGURE 1.5 Sample Test Images of Plasmodium Species.

In the severe stage of illness, parasites can be felt travelling in the blood. The analysis of illness can be made by inspecting for the parasite in a blood smear by microscope. The common procedure followed to diagnose *T. cruzi* infection is blood sample analysis from a stained thick/thin blood smear. The collected blood is examined to identify the infection rate during diagnosis. Based on the outcome, the essential treatment procedure is implemented to cure the disease. This disease is commonly found in Latin America [12]. Figure 1.6 shows the *T. cruzi* in blood smears, wherein Figure 1.6 (a) shows the adult and Figure 1.6 (b) shows the premature *T. cruzi*.

The diagnosis procedure, the advantages, and the disadvantages of the Chagas disease is similar to that of malarial infection.

1.1.2 SCREENING FOR SKIN MELANOMA

Skin cancer (melanoma) is one of the most common cancers in particularly due to high exposure of the skin to ultraviolet (UV) rays. The high exposure of pigment-producing cells called melanocytes to UV rays increases the chance of skin cancer among humans. Early diagnosis of skin cancer is essential in treating the disease.

(a)

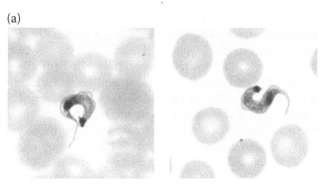

Grown Trypanosoma Cruzi

(b)

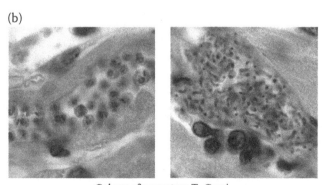

Colony of premature T. Cruzi

FIGURE 1.6 Sample Test Images of *Trypanosoma cruzi* (*T. cruzi*).

When it reaches the severe phase, melanoma will release cancer cells into the blood, thus allowing the cancer to reach other vita parts and decreasing the chance of recovery. The identification of skin cancer in its early stages needs a self check-up followed by screening procedures done by an experienced dermatologist. The dermatologist will recognise skin cancer using the commonly prescribed ABCDE rule, to distinguish the moles from melanoma.

The ABCDE rule helps to detect the skin cancer based on the following parameters:

- Asymmetric: Common moles are likely be round and balanced, but one side of the cancerous mole will probably be dissimilar to the other.
- Border: The outer surface is irregular rather than smooth and may appear tattered, uneven, or blurred.
- Colour: Skin cancer section may tend to have patchy shades and colors including black, brown, and tan.
- Diameter: Melanoma can cause a change in the size of a mole.
- Evolving: Variation in a mole's exterior over weeks or months can be a symptom of melanoma.

The clinical-level detection of skin cancer is achieved using the ABCDE rule with an examination with dermoscopy. The computerised dermoscopy will help record the suspicious skin sections for future assessment. The sample test images collected for the demonstration is presented in Figure 1.7. Figure 1.7 (a) depicts the mole (nevus), Figure 1.7 (b) presents melanoma, and Figure 1.7 (c) and (d) shows the digital dermoscopic view of the suspicious skin lesion and the skin melanoma recorded along the hair section [13]. During diagnosis, assessment along the hair section may present false result when a computerised detection is performed. Hence, in the literature, a number of skin removal systems are proposed and implemented to eliminate the skin section from the dermoscopy image. Previous research work based on skin melanoma detection under dermoscopy can be referred to in [14,15].

Skin cancer is a medical emergency in various countries, affecting a number of people each year. The clinical-level diagnosis of skin melanoma includes (i) image-assisted and (ii) biopsy-assisted detection and confirmation. In the image-assisted diagnosis, the dermoscopy images are examined by a doctor or a computer algorithm to confirm the disease with the help of the parameters presents in ABCDE rule. During the biopsy test, a sample of the skin tissue is collected and tested based on standard medical protocol. Surgery is then recommended by the doctor when cancer is confirmed during the biopsy to remove the infected skin section and to stop the cancer from spreading. Skin melanoma diagnosis with image-assisted procedures can be inferred and referred to in [15].

Advantages: Screening of skin abnormalities using dermoscopy is a widely implemented clinical practice implemented to detect skin cancer and differentiate a mole from melanoma. The availability of digital dermoscopy helps attain a digital image for further assessment.

Disadvantages: Recording of the dermoscopy needs major preparatory work when there is more skin melanoma impact. Recording of the skin section involves

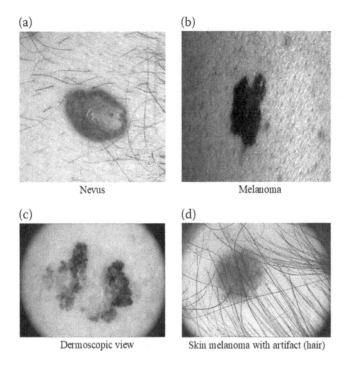

(a)

Nevus

(b)

Melanoma

(c)

Dermoscopic view

(d)

Skin melanoma with artifact (hair)

FIGURE 1.7 Dermoscopy Images to Assess the Skin Condition.

the application of a clinical gel and the removal of the hair section, which causes pain and irritation to the patient with malignant melanoma.

1.1.3 STOMACH ULCER SCREENING

Infections in the digestive tract cause Stomach Ulcer (SU), which is mainly caused by an infection in the wall of the gastrointestinal system. SU can also be caused by frequent use of drugs that affect tissues in the digestive tract. The symptoms of SU can be felt in the upper and lower gastrointestinal tract. The common symptoms of SU include continuous pain in the stomach, loss of weight, burping, vomiting blood, etc.

SU occurs when the stratum, which guards the stomach lining from stomach acid, breaks down and permits injury to the stomach coating. Infection by *Helicobacter pylori* (*H. pylori*) and the frequent usage of non-steroidal anti-inflammatory drugs also increase the impact of SU. On the onset of symptoms, the patient should immediately consult a physician to control the spread of SU.

Diagnosis and treatment of SU depends on the symptoms and severity felt by the patient. The common procedure to be followed involves:

- Barium consume: In this technique, barium in liquid form is given to the patient. This forms a lining on the upper gastrointestinal tract, which is recorded using the X-ray. The X-ray then clearly records the barium lining for better visibility, which is further assesses by the physician with a visual or computer assisted check.

- Endoscopy: Endoscopy is a common technique used to record the inner body with the help of a digital camera. The endoscopy can be inserted in with a wire, or it could be a wireless capsule. An appropriate technique then assists the diagnosis after recording the essential image, assessing the severity and orientation of SU in the gastric tract.
- Endoscopic biopsy: This technique helps record/assess the tissue in the stomach wall for tissue-level analysis. This technique involves collecting a piece of stomach tissue which would then be analyzed in the lab.

All these techniques have prominent evaluation criteria to diagnose the severity of SU in a patient. These procedures help the physician speed up the treatment process.

Gastric polyps are also a form of abnormality closely related with SU. It is an abnormal growth on the inner lining of stomach. Figure 1.8 shows a sample gastric polyps collected from the CVC-ClinicDB [16], which are collected from the frames extracted from colonoscopy videos. The research work based on this database can be referred to in [17]. This database consists of clinical-grade images along with ground-truth images provided by an expert. While performing computer-based assessment, the polyp section is extracted using a specific computer algorithm. This extracted section is then compared to the ground-truth to compute essential performance values to confirm the disease demarcation accuracy.

Advantages: The wired and wireless-capsule endoscopy is used in recording various infections both in the upper and lower digestive tract. If a clear image/video of the infection is captured, then a possible treatment can be suggested.

Disadvantages: Although rare, endoscopy can cause pain and allergy to the patient.

1.1.4 SCREENING FOR BREAST ABNORMALITY

Breast cancer is an abnormality that affects a large group of women every year and it is one of the leading cause of death [18]. Medical facilities that help women detect breast abnormalities are now accessible to them, offering preliminary screening processes. Once the nature of breast malformation is observed and

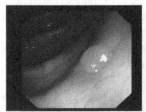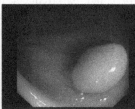

FIGURE 1.8   Stomach Abnormality Recorded Using Endoscopy.

documented, it is possible to offer appropriate treatment. Breast cancer normally develops from breast cells that form as either lobules or ducts of the breast. The initial stage of breast cancer is diagnosed through an abnormality known as Ductal Carcinoma In Situ (DCIS). These abnormal cells are usually found in the milk ducts of the breast. DCIS is known as the earliest form of the breast cancer. Clinical detection of DCIS and breast cancer is possible with imaging modalities such as Magnetic Resonance Imaging (MRI), mammogram, ultrasound imaging, and thermal imaging. The upcoming subsections present sample test images of breast cancer recorded with the above mentioned modalities and are discussed in literature found in [18,19].

Lumps found in the breast indicates the presence of irregular or abnormal breast tissue called a tumor. Mammogram and MRI are widespread radiology practices used to detect grown tumors. These techniques occasionally fail to discover breast abnormalities when the cancer is in its early stages. Hence, in recent years, thermal and sonography have been adopted to record and analyze breast malignancy due to its risk-free and contactless nature.

- **Breast MRI**

Figure 1.9 (a) to Figure 1.9 (c) presents sample breast MRI slices of the axial, coronal, and sagittal views extracted from a three-dimensional (3D) breast MRI. This image is available in the Reference Image Database to Evaluate Therapy Response (RIDER) of The Cancer Imaging Archive (TCIA) dataset [20,21]. The RIDER-TCIA which is a well-known accepted benchmark database widely adopted by researchers to test developed computerised tools. Assessment of the 3D breast MRI is computationally complex thus conversion to 2D is essential to evaluate the breast sections using a simplified computerized tool. The assessment of the 2D MRI can help accurately diagnose the tumors with better using any one or all of the views, such as axial, coronal, and sagittal. Another advantage of MRI is it records the breast section using various views. The recorded breast section can also be effectively utilized to diagnose infection level and the orientation of cancerous tumors.

(a)   (b)   (c)

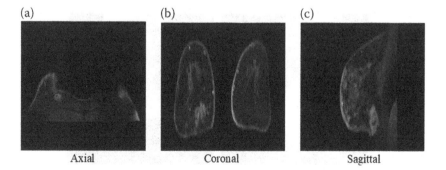

      Axial   Coronal   Sagittal

FIGURE 1.9   Breast Cancer Recorded Using MRI System.

Advantages: MRI is an imaging technique used to asses a variety of internal organs. The breast MRI is recorded using varied modalities, providing a 3D view of the abnormal breast section. The visibility is accurate in MRI compared to other imaging modalities.

Disadvantages: The assessment of the 3D MRI is complex, hence conversion to 2D is essential. It is a radiological technique that should be performed in a controlled environment under the guidance of an experienced radiologist. This acquisition modality sometimes requires the injection of a contrast agent to get better visibility of the cancerous section, which can cause several side effects.

- **Mammogram**

Mammogram another imaging technique used to record the breast section. During this imaging procedure, a specially designed X-ray source is used to record the breast section with the help of either a traditional film base or a digital technique. Due to the availability of modern X-ray systems, digital mammogram recording is now commonly used. The recorded mammogram is then examined by the doctor to detect abnormalities.

Figure 1.10 shows sample test images from the mini-MIAS database available free to public [22]. This is an accepted and used mammogram dataset to evaluate breast abnormalities [23]. It consists of a class of mammogram slices with dimensions of 1024 × 1024 × 1 pixels. It can be resized to the required size to minimize computation complexity of the image examination system employed to diagnose the breast abnormality during assessment.

Advantages: Mammogram is preferred technique for breast cancer assessment due to its low cost and simple characteristics.

Disadvantages: The recording of mammogram involves radiation and may cause mild irritation and/or pain to the breast tissues.

- **Ultrasound imaging**

For over two decades, ultrasound imaging procedures have been widely used in medical procedures to record information on the activity of internal organs, tissues,

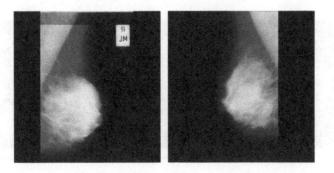

FIGURE 1.10   Breast Abnormality Recorded Using Mammogram.

and blood flow. This procedure uses high-frequency sound waves (sonography) to record essential information [24]. Compared to other imaging modalities, it is proven to be safe and causing no damage on the tissue or organ level. Recently, this imaging technique has become widely adopted for screening a variety of diseases including breast abnormalities. The ultrasound image of a breast section is depicted in Figure 1.11 presenting examples of benign (Figure 1.11 (a)) and malignant (Figure 1.11 (b)) breast cancer sections. These sections are then examined using the prescribed procedures [25].

Advantages: A simple and efficient technique to assess abnormality and activity of organs with non-invasive and simple imaging procedures.

Disadvantages: The quality of the image is very poor compared to the MRI. Ultrasound imaging requires a special diagnostic procedure detecting abnormalities.

- **Thermal imaging**

Thermal imaging is a recent imaging technique where infrared radiation (IR) is recorded using the standard protocol to construct an image of the section to be examined. Digital Infrared Thermal Imaging (DITI) is used to record the abnormal breast section for further evaluation. The level of IR wave coming out of a body organ mainly depends on its condition. The thermal camera captures the radiation and converts it into image patterns using a dedicated software unit. The recorded thermal image characterizes an image pattern based on the level of the IR wave. It is possible to detect the abnormality by analysing the recorded patterns from the image.

Figure 1.12 presents thermal images recorded using the prescribed imaging modality available in [26]. Figure 1.12 (a) and (b) depicts the greyscale version of thermal imaging while Figure 1.12 (c) and (d) depicts the RGB scale version. Figure 1.12 (e) and (f) depicts the gray and RGB scale image of a chosen patient with the breast abnormality. These images confirm that the thermal imaging modality is a non-invasive technique considered to record the essential images from the infected person and can be used to diagnose the abnormality. The earlier work

(a) (b)

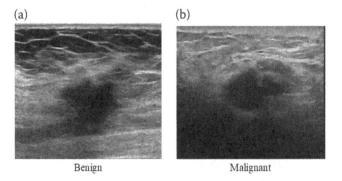

Benign Malignant

FIGURE 1.11 Breast Cancer Recorded Using Ultrasound Technique.

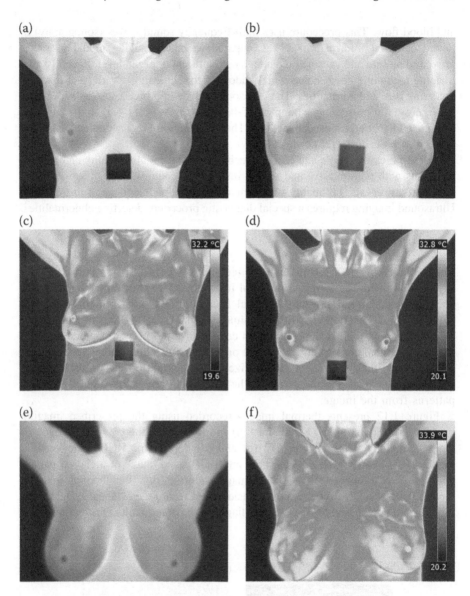

FIGURE 1.12 Breast Abnormalities Recorded Using Thermal Imaging technique.

on breast thermogram analysis from the DCIS for cancer detection can be found in [18,19,27,28].

Advantages: It is a recent and non-contact imaging system widely adopted by hospitals and scan centers. It is a handy system where a special digital camera is used to detect and convert the IR radiation into images.

Disadvantages: The assessment of the thermal imaging needs special procedures since it is constructed using thermal patterns.

- **Histology**

Histology is a branch of study where the microanatomy of various cells, tissues, and organs are examined using digital microscopic images. This technique inspects the association among arrangement and functional difference between the healthy and disease samples collected from the patients. Normally, the histology based assessment is used to confirm the condition of the disease using biopsy samples [29].

To confirm breast cancer, a needle biopsy is collects breast tissue and then it is examined using the digital microscopy. Figure 1.13 presents the histology images of the breast cancer samples, in which Figure 1.13 (a) shows the benign and (b) shows the malignant class. The histology analysis normally offers a better diagnostic vision regarding breast cancer evaluation and treatment.

Advantages: A tissue-level diagnosis of the disease is possible, helping design accurate treatment.

Disadvantages: This technique requires tissue collection through needle biopsy which can be a painful procedure. The preparation of histopathological slides needs special care. Diagnosis is possible with an experienced doctor only.

This sub-section presented various imaging modalities available to assess breast abnormalities. Each modality has its own merits and demerits, and the choice procedure depends on the doctor's expertise. If the cancer is less visible in image-based analysis, the doctor may suggest a biopsy-assisted examination to confirm the condition (benign/malignant) of the cancer. This helps design the essential treatment procedure. Imaging techniques such as breast thermogram and sonography appear to be harmless techniques compared to the mammogram and breast MRI.

1.1.5 SCREENING FOR BRAIN ABNORMALITY

The brain is the primary organ responsible for appraising physiological signals coming from other sensory sections. It is also in charge of taking necessary

(a)   (b)

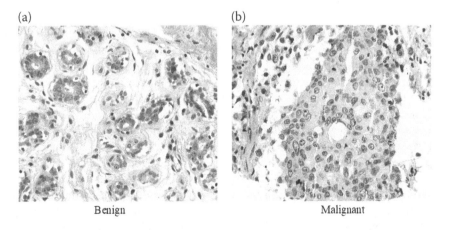

Benign   Malignant

FIGURE 1.13 Microscopic View of the Breast Tissues with Cancer Cells.

managing actions. This process is affected negatively if any illness or disease develops in the brain. Unobserved and unprocessed brain sickness can cause various problems, including death. The standard state of the brain may be affected by different causes such as birth defects, head wound by an accident, and cell growth in the brain segment. A substantial amount of alertness programs and anticipatory measures are being taken to protect the brain from malformation, but due to factors such as current lifestyle, food behavior, genetics and age there is rising distress over brain defects. If the brain defect is detected in its early stages, then measures can be implemented to control the illness. The brain normally has a large quantity of soft tissues along with the connected signal transceivers, hence biopsy is not recommended to diagnose these abnormalities [30–33].

The condition of the brain is assessed by bio-signals collected using standard electrode system or by the bio-images collected using standard imaging modalities. The signal (EEG) and image (MRI or CT) based assessment offer information regarding a considerable number of the brain diseases as discussed by Rajinikanth et al. [33]. Evaluation of brain defects with EEG is moderately complex, thus requiring different pre-processing and post-processing methods. Hence, image-assisted techniques are performed in most clinical-level diagnosis. This sub-section discusses a variety of procedures considered to examine the brain condition.

- **EEG-Assisted Screening**

The EEG-assisted brain evaluation is a common procedure followed in clinical assessment for various abnormalities that develop in the central nervous system. A standard signal acquisition procedure is considered to record bio-electric potentials from the brain which is done using a predefined scalp electrode setup. The assessment of the EEG is quite complex compared to the image-assisted technique due to its non-linearity. Hence, in recent years, the recorded EEG signals are converted into equivalent images using a signal-to-image conversion technique. The converted images are then effectively examined using image processing procedures.

In this sub-section, the benchmark EEG signal of Bern-Barcelona-Dataset (BBD) [34] is considered for epilepsy assessment. This dataset consists of 100 (50 focal + 50 non-focal) one-dimensional (1D) EEG signals recorded using the clinically admitted protocol. This dataset is widely used in research due to its complex and non-linear nature. The major difference in the focal/non-focal (F/NF) EEG is the change in its amplitude level and the frequency pattern. Figure 1.14 presents the sample EEG signal of BBD; Figure 1.14 (a) and (b) shows the patterns of focal and non-focal EEG signals.

The focal and non-focal EEG depicted in Figure 1.14 have particular amplitudes and frequency patterns. The assessment of amplitude and frequency pattern can be achieved using a statistical and signal evaluation technique to identify brain abnormality. To simplify the assessment, the Gramian-Angular-Summation-Field (GASF) technique is implemented to convert the EEG signal into an RGB image with different patterns, which can then be examined with an imaging procedure to identify the condition of the brain [35].

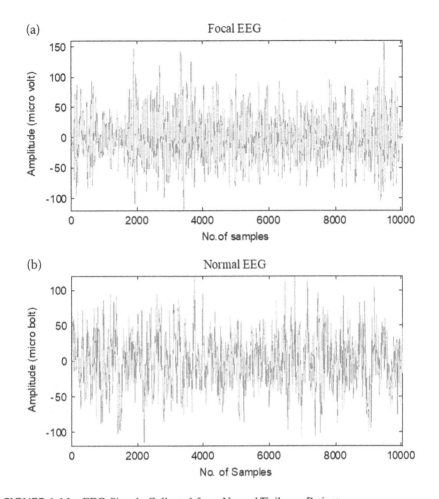

FIGURE 1.14   EEG Signals Collected from Normal/Epilepsy Patients.

The conversion of the EEG into RGB pattern using the GASF is presented below:

In this conversion, the total length of EEG signals (10,240 samples) is divided into eight equal segments (10240/8 = 1280 samples) and the segmented signal is then converted into images with dimensions of 256 × 256 × 3 pixels. This process generates 800 (400 + 400) GAF patterns for focal and non-focal EEG signals of BBD. The segmentation procedure executed on the focal class of EEG is shown in Figure 1.15.

The GAF was initially proposed by Wang and Oates [36] and, due to its practical significance, it has become widely adopted by researchers to analyze bio-signals.

The mathematical description of GAF is as follows:

Let, $T = \{t_1, t_2, ..., t_n\}$ represent the time series of a finite signal. Then the normalization of the series ($T \in [-1, 1]$) can be achieved using

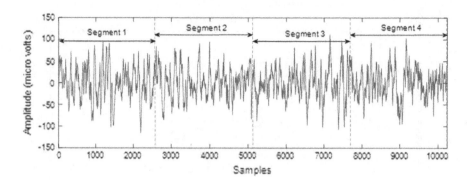

FIGURE 1.15   Sample GASF Implementation for an EEG.

$$\overline{T}_{-1}^{-i} = \frac{(t_i - \ max(T)) + (t_i - \ min(T))}{max(T) - \ min(T)} \qquad (1.1)$$

When the rescaled time series (T') is transferred into polar coordinates (ϕ), it can be represented as

$$\begin{cases} \phi = arccos(t'_i), \ -1 \le t'_i \le 1, \ t'_i \in T' \\ r = \frac{x_i}{N}, \ x_i \in N \end{cases} \qquad (1.2)$$

where x_i = time stamp and N = constant feature to normalize the duration of polar coordinate.

The final GAF is generated using the following equation:

$$GAF = \begin{bmatrix} \cos(\phi_1, \ \phi_1) & \cdots & \cdots & \cos(\phi_1, \ \phi_n) \\ \cos(\phi_2, \ \phi_1) & \cdots & \cdots & \cos(\phi_2, \ \phi_2) \\ \cdot & \cdots & \cdots & \cdot \\ \cos(\phi_{256}, \ \phi_1) & \cdots & \cdots & \cos(\phi_{256}, \ \phi_{256}) \end{bmatrix} \qquad (1.3)$$

If equation (1.3) is implemented, then the EEG with 1280 samples is converted into an image. The sample test images of F and NF class obtained using GASF can be visualized from Figure 1.16. The generated GASF patterns are used to evaluate brain abnormalities. Figure 1.16 (a) depicts the focal EEG pattern in which large or more visible pixels are observed when compared to the non-focal class. From Figure 1.16 (b), it could be noted that the pattern developed by the normal (non-focal) EEG is simple where a repetitive pattern is observed. A developed computer algorithm or a visual observation is necessary to identify the type of EEG signal based on the variation in the pattern. Previous research work carried out with this EEG signal and GASF based assessment can be found in [35].

(a)

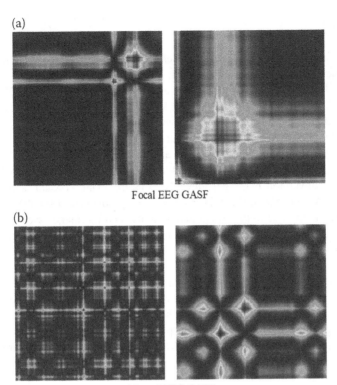

Focal EEG GASF

(b)

Normal EEG GASF

FIGURE 1.16 Patterns Made by the Normal/Epilepsy EEG Signals Converted Using GASF Technique.

Advantages: The electrical activity of the brain can be easily diagnosed using the EEG, the extraction which can be achieved with a single or multi-channel EEG system. The EEG signal can be collected using surface electrodes placed on a predefined place adopting the 20-20 electrode system.

Disadvantages: This signal pattern is very non-linear and in most cases it will vary based on the induced noise and various activities such as thinking, speaking, etc. The assessment of the EEG signal needs a special detection procedure and is quite complex and less reliable compared to the image-based techniques.

- **Image-Assisted Screening**

Apart from the signal based assessment of the brain, various image-based detection techniques are also available these days. The assessment of brain abnormalities by imaging modalities is quite simple due to better visibility and a number of recording tools available. This sub-section provides information on the assessment of brain tumor and stroke using MRI and Computed Tomography (CT) recorded images.

- **MRI**

The assessment of brain abnormality is very commonly performed using the MRI due to its varied modalities such as Flair, T1, T1C, T2 and DW [32]. Figure 1.17 shows T2 modality of brain tumor images, such as Glioblastoma Multiforme (GBM) and the Low Grade Glioma (LGG) from the TCIA database [37]. This dataset offers clinical-grade MRI images collected from the volunteers using a prescribed protocol. The published work based on this dataset can be found in [30,31].

These dataset images are in 3D form and, to minimize the complexity during the assesment, a 3D to 2D conversion is achieved using the well-known ITK-Snap tool [38]. This conversion helps get views such as the axial, coronal, and sagittal orientations. Among them, the axial view is widely considered for the assesment of brain abnormality due to its structural simplicity. During image assesment, the skull section can be removed using a standard skull stripping technique after which a computer algorithm can be executed for the automated assessment of the tumor section in the images under consideration.

During the recording of the MRI, a contrast agent called the Gadolinium is injected into the patient. This helps in recording the tumor section by providing better contrast compared to other brain sections. In Figure 1.17, the GBM and the LGG images are presented for the purpose of assesment. It is observed that the LGG MRI slice has less tumor dimension compared to the GBM. Similar to brain tumor detection, MRI modality is also considered for the assessment of ischemic stroke lesion.

Figure 1.18 depicts the MRI images of the benchmark stroke dataset called the Ischemic Stroke Lesion Segmentation Challenge (ISLES) available in [39].

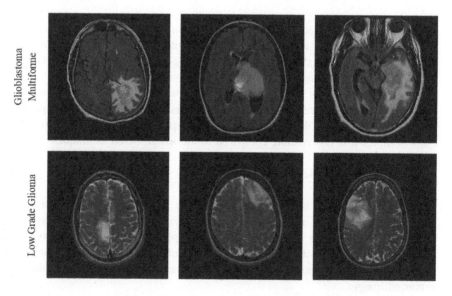

FIGURE 1.17 Assessment of Brain Tumour with MRI Slices in T2 Modality.

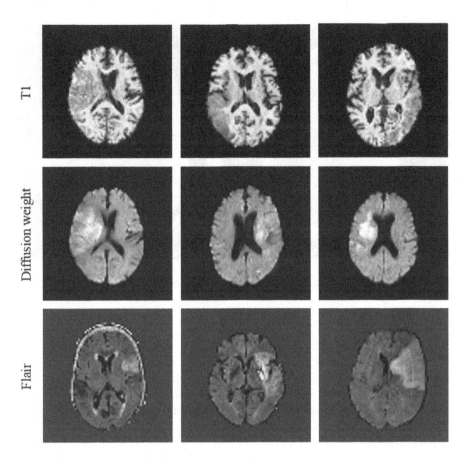

FIGURE 1.18 Ischemic Stroke Assessment of MRI Slices.

The MRI slices of T1, DW and the Flair modalities are presented in the Figure 1.18. Based on the requirement, a single modality image could be considered for the assesment. It is observed from Figure 1.18 that the stroke lesion is more visible in DW and Flair modality of MRI when compared with T1. Hence, the assesment of the T1 image is quite difficult compared to the alternatives. Research work carried out based on the ISLES assesment could can be accessed from [40–43].

Advantages: Brain MRI is commonly used to record brain abnormalities such as stroke and tumors with the help of a reconstructed 3D image. The visibility of the brain abnormality is accurate compared to other imagining techniques. Further, the brain MRI supports different modalities such as T1, T1C, T2, Flair and DW.

Disadvantages: The assessment of 3D brain MRI is complex, hence a 3D to 2D conversion is required to examine the brain abnormality using the axial, coronal and sagittal views. Imaging occasionally needs the injection of a contrast agent to capture improved visibility of cancerous segment. This agent may induce temporary or even permanent side effects in patients.

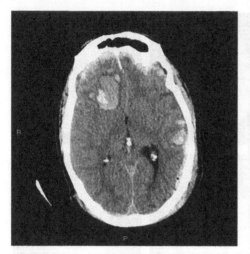 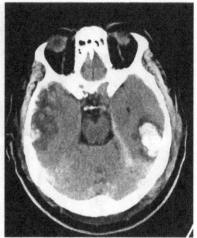

FIGURE 1.19   CT Scan Slices of Brain with Tumour Section.

- **CT scan**

Computed Tomography (CT) imaging is also widely used to record brain images for the assessment of abnormalities in the brain. Figure 1.19 presents the CT scan slices with brain tumor sections collected from Radiopaedia [44,45]. Similar to MRI, CT scan is also a radiology-assisted imaging procedure widely considered to assess abnormalities in the brain with better accuracy.

Advantages: The CT is a simple and commonly adopted radiological technique implemented to record the structure and abnormality of internal organs along with the bone section. It is a low-cost imaging procedure.

Disadvantages: Assessment of the abnormality using CT is quite difficult when compared with MRI because image information in CT scan slice is less visible.

1.1.6 Screening for the Fetal Growth

Prenatal screening is a procedure for prenatal diagnosis achieved with obstetric ultrasonography or prenatal ultrasound. This procedure helps detect the fetal growth during pregnancy using ultrasound-assisted imaging. After confirming the pregnancy, continuous and scheduled testing is essential to identify abnormalities in fetus growth. Carrying out an ultrasound examination at the early stages of pregnancy can assist the precise analysis of the onset of pregnancy. It also aids in finding multiple fetuses and major congenital abnormalities. The common procedure followed in prenatal diagnosis is to predict fetus growth by simply measuring the head circumference. Previous work on prenatal screening can be found in [46] and the images collected from the benchmark HC18 database [47] is presented in Figure 1.20.

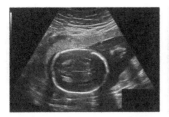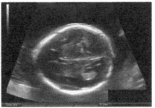

FIGURE 1.20 Ultrasound Imaging-Based Fetal Growth Assessment.

Advantages: It is a simple and cost-effective imaging procedure used to record and analyze fetal growth. This technique provides accurate information regarding the movement of the fetus and is one of the safest imaging modalities.

Disadvantages: This method results in blurred images which can only be examined by an experienced doctor.

1.1.7 SCREENING FOR RETINAL ABNORMALITY

Due to various unavoidable causes such as aging, eye diseases develop irrespective of race or gender. A disease of the eye usually requires a personal check-up by an experienced opthalmologist followed by image-guided assesment. During image-assisted diagnosis, a special imaging system, called the fundus camera, is used to record the retinal section for further assesment. Fundus cameras record the retina, the neuro-sensory tissue in our eyes that decodes the optical images into an electrical impulses. While the retinal images are being recorded, the patient is advised to sit in front of the fundus camera with their chin placed on a chin rest and their forehead against a bar. The opthalmologist then focuses and aligns the fundus camera. A light ray is passed into the eye through which the essential image is recorded [48,49]. Existing literature confirm that the fundus image-assisted technique can be used to detect a variety of eye diseases. Figure 1.21 depicts a typical fundus image collected from the benchmark retinal database present in literature. Figure 1.21 (a) to (d) presents the images of normal, Age-Related Macular Degeneration (AMD), diabetic retinopathy, and macular edema. This retinal image is then assessed by an experienced ophthalmologist or a computer algorithm to assess the abnormality. Further, this information is considered to plan and execute procedures for treatment and follow-up.

Advantages: It gives a clear picture of the retina using an RGB-scaled image recorded with a fundus camera. It is one of the common procedures to record and examine eye-related abnormalities.

Disadvantages: The information available in the image can be examined only by an ophthalmologist. Occasionally it fails to record the vital information and requires the correction positioning of eye.

1.1.8 SCREENING FOR LUNG ABNORMALITY

The lung is the vital organ responsible for supplying oxygen to the human body. The purpose of the respiratory arrangement is to extract oxygen from the

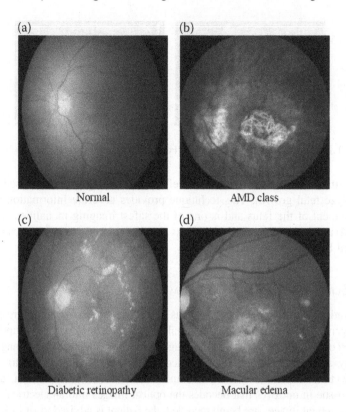

FIGURE 1.21   Sample Retinal Images Collected by Fundus-Camera.

atmosphere and shift it into the bloodstream while discharging carbon dioxide from the bloodstream into the atmosphere, a process of gas exchange. Humans have two lungs which are located in the thoracic cavity of the chest. The right lung is bigger than the left, which shares space in the chest along with the heart. The tissue of the lungs can be infected a variety of respiratory illnesses including pneumonia and lung cancer. The recently-emerged COVID-19 also infects the lungs at a considerable rate, causing medium/heavy pneumonia. If left untreated, lung disease can affect the respiratory system and disturb the gas exchange between the bloodstream and the atmosphere [3].

- **Chest Radiograph**

A chest radiograph (chest X-ray or chest film) is normally used to identify several conditions linking the chest wall, lungs, heart, and great vessels. Literature review confirms that pneumonia and congestive heart malfunction are normally diagnosed using chest X-ray. Normally, the chest X-ray is suitable for recording and evaluating the normal/abnormal conditions of the chest. It provides a good screening output for further diagnosis. This sub-section describes detailed information regarding pneumonia and tuberculosis identification from chest X-ray collected from the benchmark datasets [50,51].

- **Pneumonia diagnosis**

Pneumonia is caused by an infection in the respiratory tract due to the microorganisms such as bacteria, virus, and fungi. Pneumonia causes a range of abnormalities in respiratory system, preventing oxygen exchange to the bloodstream. Untreated pneumonia is acute for the children aged five years and below and the elderly aged 65 and above. To provide appropriate treatment for those affected, it is essential to diagnose the infection rate in the lung. Due to its clinical significance, a considerable number of semi-automated and automated disease diagnosis systems are proposed and implemented to diagnose various lung abnormality using the chest X-ray.

Figure 1.22 depicts the chest X-ray collected from the available pneumonia public database [50,51]. Figure 1.22 (a) to (c) presents chest radiograph of the normal, bacterial and viral pneumonia cases respectively. This radiograph slides are then examined using appropriate technique to assess the abnormality.

- **Tuberculosis diagnosis**

Tuberculosis (TB) is an acute lung disorder caused by the Mycobacterium tuberculosis bacteria. TB normally infects the lung and causes acute respiratory distress. In some special cases, TB infection can infect other body organs in what is known as extra-pulmonary TB. The assessment of TB is normally done using (i) imaging technique, (ii) endoscopy-assisted procedure, and (iii) needle biopsy.

The image-assisted methodology based on chest X-ray or lung CT is the most commonly used modality to assess TB. Chest X-ray can be used for initial diagnosis, but due to its poor image interpretation, CT images are preferred to convey useful information. Figure 1.23 depicts the sample chest X-ray of a TB case collected from Radiopaedia [52,53].

(a)　　　　　　　　(b)　　　　　　　　(c)

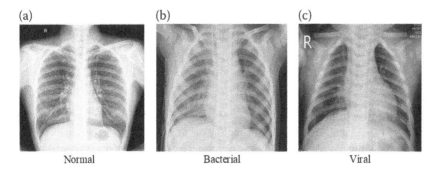

Normal　　　　　　　Bacterial　　　　　　　Viral

FIGURE 1.22　Chest Radiographs Collected from Normal and Abnormal Patients.

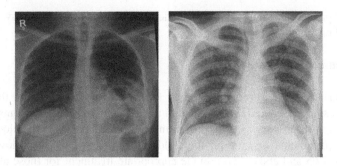

FIGURE 1.23   Chest Radiographs of Normal and Patients with Tuberculosis.

Advantages: Chest radiograph is a simple and commonly implemented imaging modality to examine infections of the lung.

Disadvantages: It results in less visibility of the infected section compared to other imaging techniques such as CT scan.

- **Lung CT-Scan**

Computed Tomography (CT) scan is a widely adopted imaging technique in hospitals to assess a variety of organs including lungs. The main advantage of CT is that it offers a reconstructed 3D image which can be examined either in 3D or 2D form. The assessment of the CT helps to identify the disease accurately when compared to the chest X-ray. It can be noted from the literature that the CT scan slices of axial, coronal, and sagittal views are used in assessing abnormalities, and the choice of a particular orientation is decided by the physician [54–57].

- **Pneumonia**

The CT scan assisted pneumonia diagnosis is a widely adopted technique in health centres due to its proper diagnostic capability. Figure 1.24 depicts the sample test images acquired from the pneumonia infected patients. The highly visible section is due to the pneumonia lesion and can be extracted and evaluated during the assessment. Based on the infection level, the treatment process and follow up of the patients is decided by the pulmonologists.

- **Tuberculosis**

The CT scan is considered to examine pulmonary TB infection in lungs with better accuracy. Figure 1.25 presents information on lung infection existing in the axial and coronal views of CT scan slices. Compared to the chest X-ray, infection due to the TB is more visible in lung CT scan slices thus making it easier to be detected and interpreted by the technician or computer.

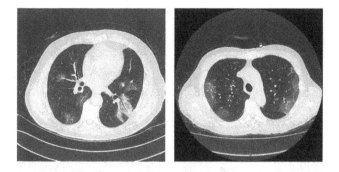

FIGURE 1.24   Lung CT Scan Slices (Axial View) Acquired from Patients with Pneumonia.

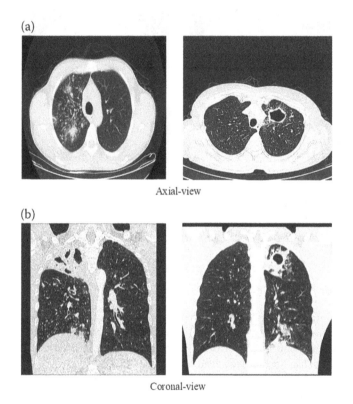

FIGURE 1.25   Lung CT Scan Slices Collected from Patients with Tuberculosis.

- **Lung Nodule**

A lung nodule is an abnormality that develops mostly from damaged cells due to cancer. There are several image examination techniques present in the literature to examine lung nodule so as to assess the different stages of disease severity. Figure 1.26 presents 2D CT scan slices of benign and malignant lung

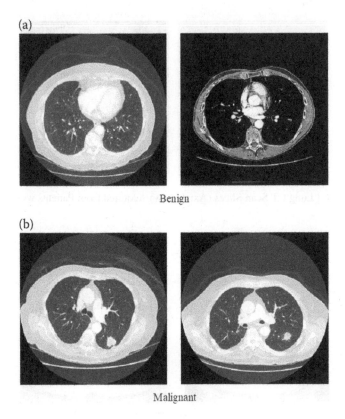

FIGURE 1.26 Lung CT Scan Slices (Axial View) Depicts Pulmonary Nodules.

nodule images collected from Lung Image Database Consortium (LIDC-IDRI) dataset [58].

1.1.9 HEART DISEASE SCREENING

The heart is another vital organ in the body. Abnormalities in the heart can affect normal human activities as the heart is responsible for blood circulation. Heart abnormalities develop due to reasons ranging from the infection due to aging to untreated heart disease that may lead to death.

The functioning of the heart can be observed using signal-(ECG) based and image-guided assessment. The image-guided examination provides essential visual information regarding the heart. Assessment of images is quite straightforward compared to signals. Figure 1.27 shows heart MRI slices of the benchmark HVSMR 2016 challenge database referred to in [59]. This dataset is available along with its GT image but the task is to segment a particular section from the MRI slice with better accuracy. Once segmentation is completed, comparison against the GT is done. Performance of the proposed technique is validated by computing essential performance measures. The previous literature work based on this database can be accessed in [60].

1.1.10 Osteoporosis

Bones are a major components in human physiology that provide its shape and strength. Normally, it is formed by calcium minerals salts. A predefined bonding is executed with collagen fibres to get the definite shape and strength of the bone. All bone structures have an active tissue that continually repairs itself based on the need but this activeness decreases due to a variety of factors including age. In children and adults, new bone structure forms quickly compared to the elderly. Further, bone density also varies based on age and calcium level. One of the abnormalities related to the bone is osteoporosis, which causes a change in the physical structure of the bone leading to bones becoming weak and brittle due to vitamin and calcium deficiency. It is a common affliction of the elderly who don't take sufficient minerals (calcium) required for their age. Previous literature confirms that osteoporosis affects more women compared to men. Early detection and proper diet can reduce the impact of the disease. Detection procedure involves recording and analysing the microstructures of the bone using an image-guided assessment. Figure 1.28 illustrates various views of the bone section using CT scan modality especially for bone disease assessment. Information regarding the presented images can be found in Radiopaedia database referred to in [61–64].

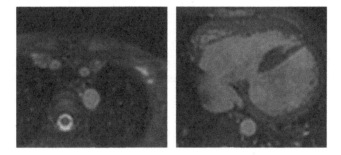

FIGURE 1.27 Heart MRI Slices.

(a) (b) (c)

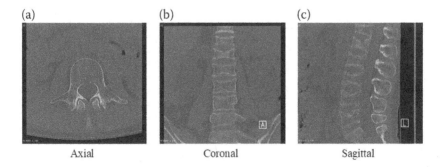

Axial   Coronal   Sagittal

FIGURE 1.28 CT Scan Slices for Osteoporosis Assessment.

Advantages: Examination of osteoporosis using the CT scan slice is more informative compared to other techniques.

Disadvantages: The information available regarding the bone and its internal structure is incomplete as there are other vital information essential to examining the bone structure with appropriate accuracy.

1.1.11 SCREENING OF COVID-19 INFECTION

Recently emerged communicable Coronavirus Disease (COVID-19) is now becoming a global threat. It has infected a large number of people worldwide due to its rapid spread. COVID-19 affects the respiratory tract, causing the acute pneumonia. The symptoms of COVID-19 depend on the patient's immune system. Common symptoms vary from dry cough to difficulty in breathing.

The Disease Screening (DS) for COVID-19 consist of two phases: (i) Reverse Transcription-Polymerase Chain Reaction (RT-PCR) test, and (ii) image-assisted diagnosis to confirm the disease. RT-PCR is a laboratory-level detection process performed using samples collected from the infected person. If a positive RT-PCR test is encountered, then the doctor can suggest an image-assisted diagnostic to confirm the disease and its severity level. During the image-assisted diagnosis, the infection in the lung is recorded using Computed-Tomography (CT) scan images and/or chest radiographs (Chest X-Ray). The recorded images are then examined by an experienced doctor and, based on his/her observation, treatment and observation to further follow up is suggested. Figure 1.29 shows chest X-ray of the infected patient and Figure 1.30 shows the CT scan slice. The infection severity is examined by a radiologist and follow-up is done by a doctor. Based on this outcome, a suitable treatment is planned and implemented by the physician to alleviate the problem.

1.2 MEDICAL IMAGE RECORDING PROCEDURES

Many researchers have reported on the prognosis and diagnosis of diseases, confirming that signal/image-assisted DS procedures are implemented globally. Non-invasive procedures are preferred over invasive methods due to its simplicity and

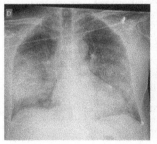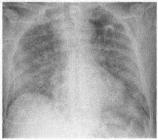

FIGURE 1.29   Chest X-Ray Acquired from the Patient with COVID-19 Infection.

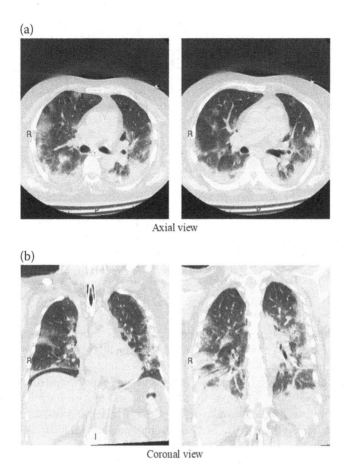

(a)

Axial view

(b)

Coronal view

FIGURE 1.30 Lung CT Scan Slices Acquired from the COVID-19 Patient.

safety. Further, diagnosis based on images is preferred compared to the diagnosis arrived at through signals. The developments in digital imaging systems have paved a way for implementing image-assisted disease detection procedure in most of disease cases.

Normally, medical images are recorded in a controlled environment by a standard imaging procedure. Each imaging procedure has its own protocol that is executed by a skilled technician under the supervision of a doctor if necessary. Medical images recorded under a certain approach vary based on the abnormality to be detected and the organ to be examined. Mostly, diseases in external organs are initially examined via a personal check by an expert. Based on the initial diagnosis a standard image acquisition protocol is suggested for further assessment. Detection of an abnormality in an internal organ, on the other hand, is complex and requires a series of identification procedures that are executed in a controlled environment with supervision by a team of experts. Figure 1.31 depicts commonly followed imaging methods and the relevant internal/external organs. The outcome of this

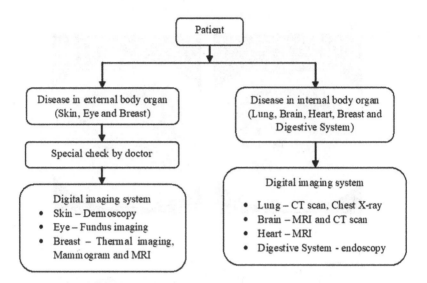

FIGURE 1.31   Image-Guided Diagnosis of Disease.

process is then verified by the field expert (doctor) to decide the treatment procedure to be implemented to control and cure the abnormalities.

Medical imaging techniques are common procedures executed to assess diseases in various parts of the body. The imaging practice is implemented based on the need to record various sections such as the cell, tissue and organs. Each imaging methodology has its own procedure and would help to provide the medical images with (i) varied color (Greyscale/RGB), (ii) different dimension (2D/3D) and (iii) varied orientations.

The choice of a particular image recording methodology mainly depends on the disease to be diagnosed and the choice of the doctor, who decides the flow of the procedure based on his expertise. Each procedure is performed in a controlled environment with the supervision of a lab technician or an experienced doctor. The recorded images can be examined in imaging centers, hospitals, and in both locations based on the disease to be diagnosed and the treatment to be executed.

This section presents the recording procedures commonly implemented: (i) blood screening, (ii) skin cancer diagnosis, (iii) stomach ulcer detection, (iv) detection of breast cancer, (v) brain abnormality detection with MRI, (vi) fetal growth scanning, (vii) retinal abnormality assessment, (viii) lungs infection detection, (ix) osteoporosis and (x) COVID-19 screening.

- **Blood screening**

Blood screening is initiated by collecting a required volume of blood from the patient whose condition is to be screened. After collecting blood, the screening procedure begins with the help of contrast-enhanced thin/thick blood films. The prepared blood film is then examined under a conventional or a digital microscope.

Screening of the blood is a common procedure in hospitals and clinics, helping diagnose the initial condition of the patient with a considerable number of measurements from WBC to the fat and sugar level in the blood. The common diagnostic stages involved in blood sample evaluation are presented in Figure 1.32. This procedure results in a microscopic image for further examination.

- **Skin cancer diagnosis**

Skin cancer a major diseases seen in the elderly and those who are exposed to high UV radiation. The common procedure followed in diagnosing the skin level infection using dermoscopy is presented below. If the imaging procedure presents a suspicious result, then biopsy is implemented to confirm the disease. If skin cancer is confirmed, then treatment is designed to control the spread of the cancer to other tissues. The infected skin section is carefully removed through surgery (Figure 1.33).

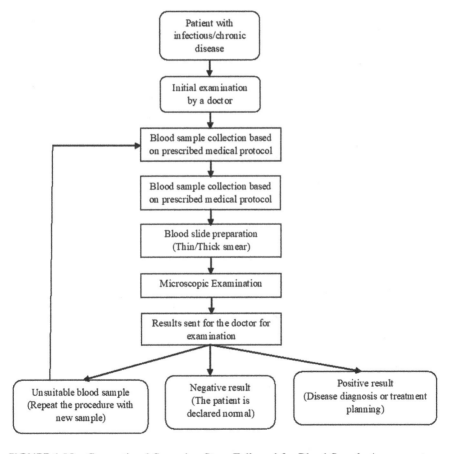

FIGURE 1.32 Conventional Screening Steps Followed for Blood Sample Assessment.

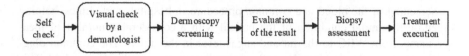

FIGURE 1.33   Diagnostic Stages Implemented to Detect Skin Cancer.

- **Stomach ulcer detection**

Stomach ulcer is an illness that develops due to a variety of reasons including uneven or poor diet. This disease is identified with appropriate care as, if left untreated, it may lead to stomach cancer. Normally, the infection rate is diagnosed with the help of a wired or a wireless endoscope, which provides a clear picture or a video of the stomach wall (Figure 1.34).

- **Breast cancer detection**

Breast cancer develops in women due to factors such as aging. The stage and severity of the disease for a person suffering from a developed breast cancer can be diagnosed through clinical biopsy, which helps examine the breast tissues and cells. Further, the image-assisted modalities such as MRI, ultrasound, mammogram, and thermal imaging are also used to examine the breast abnormality with better diagnostic accuracy. Compared to the needle biopsy, imaging techniques are noninvasive procedures done by visual examination of a physician or performed with the help of computer algorithm (Figure 1.35).

- **Brain abnormality examination**

The brain is the major decision-making organ controlling other parts of the body based on what is needed. Abnormalities in the brain lead to conditions such as the behavior, speech, walking pattern, and quick decision making problems. Untreated brain abnormalities may lead to temporary and even permanent disability and, in certain cases, in mortality.

Problematic conditions in the brain are usually assessed using the brain signals and image-assisted tools. Clinical-level assessment of signal/image-assisted monitoring is essential. Figure 1.36 depicts the conventional procedures followed while the activity of the brain is being recorded. Image-assisted techniques such as MRI and CT are widely preferred due to the simplicity in which the brain condition is diagnosed. These images are examined by a radiologist and a doctor and their combined decision is considered for designing treatment. The choice of a (image/ signal) procedure depends mainly on the physician.

- **Fetal growth scanning**

It is essential to have routine health checkups to avoid future maternal complications once pregnancy has been confirmed. An ultrasound-assisted scan is

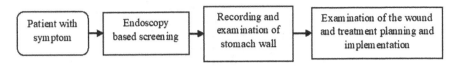

FIGURE 1.34   Stomach Ulcer Detection Based on the Endoscopic Imaging.

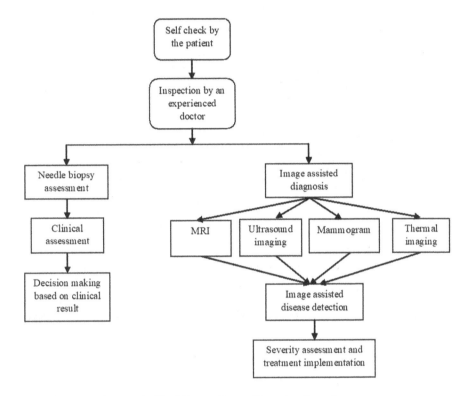

FIGURE 1.35   Commonly Used Breast Cancer Detection Procedure.

recommended to monitor fetal growth between predefined periods (e.g., 28 to 32 weeks). This scan can be done several times based on the suggestion of the doctor (Figure 1.37).

During this process, a clinically-accepted ultrasound scanner (transmitted and receiver) is used to record the orientation and activities of the fetus. Also, continuous scanning can help measure the rate of fetal growth. The image attained from scanning can be monitored using a display unit or a recording on a transparent film. The assessment will help identify birth defects and other complications at an early stage and, if needed, suggestions can be given by the doctor to mainly monitor the growth in the womb.

- **Retinal abnormality examination**

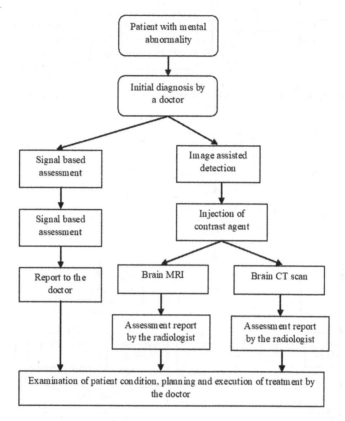

FIGURE 1.36   Clinical Level Diagnosis of Brain Abnormality with Signal/Image-Based Techniques.

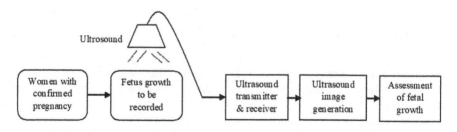

FIGURE 1.37   Evaluation of Fetus Growth with Ultrasound Imaging.

The retina is a major sensory organ which converts light signals into meaningful information/picture for the brain. Disease in the eye severely affects the sensing and the decision-making process of the brain. Eye abnormalities can be caused by diabetes, infection, aging, accident etc.

FIGURE 1.38 Screening Practice for Eye Abnormalities.

Eye abnormalities due to diabetes and aging are of primary concern. There is a considerable number of dedicated screening procedures that could be carried out in clinics. The fundus camera based retina examination is one such common procedure. The process followed for the retinal examination is described in Figure 1.38. After getting the essential image, a visual check or a computerised analysis is implemented for identification of retinal diseases.

- **Lung infection detection**

Abnormalities in the lung can severely affect the respiratory system and disturb normal air exchange. Oxygen deficiency in the bloodstream can affect normal activities of other parts, cell movements are disturbed. If this problem persists, the body is negatively affected, which may lead to organ failure or death.

Abnormalities in the lungs is diagnosed through a CT scan or a chest X-ray as it is covered by a complex bone section called the ribs. The clinical-based imaging system used in radiology to record the activity and condition of the lungs is presented in Figure 1.39. In this procedure, the 2D or 3D version of the image is recorded and is then assessed by a physician or a computer algorithm to accurately detect the abnormality.

- **Osteoporosis diagnosis**

The bone's weakening heightens the risk of breakage, which is a common outcome of osteoporosis. Common bones that usually break up are the vertebrae, the bones in forearm, and the hips. Osteoporosis can be prevented by appropriate medication and diet when it is identified in its early stages. The procedures followed during osteoporosis screening are illustrated in Figure 1.40. Normally, CT scan slices recorded in a controlled environment are used to detect the deficiency and its severity.

- **COVID-19 screening**

Recently, the infection caused by the novel SARS-CoV-2 (COVID-19) virus was declared as a pandemic by the World Health Organisation (WHO). This virus has caused a global health emergency. It can cause mild to severe infection in the respiratory tract, but timely screening and isolation of the disease can help control its spread. Screening for COVID-19 infection should follow WHO guidelines. It involves sample-assisted testing with the RT-PCR followed by an image-assisted diagnosis.

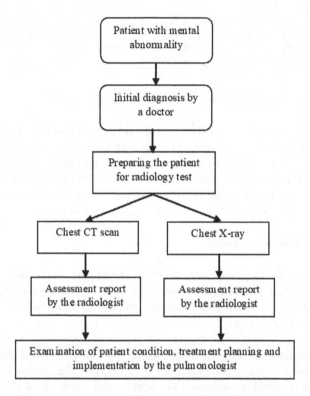

FIGURE 1.39   Clinical Level Screening of the Lung Abnormalities.

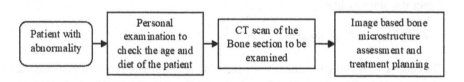

FIGURE 1.40   Detection and Evaluation of Osteoporosis.

The clinical procedures followed to detect COVID-19 infection are presented in Figure 1.41. Clinical-level testing is initiated based on the collected samples from the patient when symptoms for COVID-19 are present. If the initial RT-PCR test result is negative, the patient is declared healthy and is advised to self-isolate for a predefined time. If RT-PCR rest is positive, then a suitable imaging procedure based on the Chest X-ray or CT scan is executed to identify the infection rate. Based on this information, treatment is initiated. The image-assisted technique in COVID-19 detection is a commonly adopted procedure in hospitals. This helps evaluate the infection rate before treatment is implemented. Previous works on the image-assisted diagnosis of COVID-19 can be referred to in [65–70].

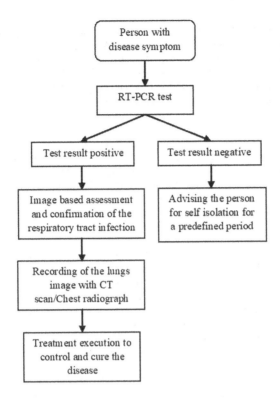

FIGURE 1.41   Screening Procedures Followed in COVID-19 Disease Diagnosis.

1.3 SUMMARY

This chapter discussed information on various diseases and the imaging modalities available to assess diseases in various organs. For every organ, an appropriate imaging procedure is implemented to capture an image with better visibility under a controlled medical environment. Further, this section presented various procedures implemented to record and examine the images. From this section, it can be noted that a number of imaging techniques are available for the disease screening process for each organ where the choice of a particular procedure depends mainly on the expertise of the physician. The recorded images can be converted digitally and examined using a computer algorithm. This procedure aids the doctor to reduce the diagnostic burden. The outcome of the computerized technique as well as the observation by the doctor together can result in efficient planning of treatment.

REFERENCES

1. Chaki, J. & N Dey, N. (2020). Data tagging in medical images: A survey of the state-of-art. *Current Medical Imaging*. doi: 10.2174/1573405616666200218130043.
2. Dougherty, E.R. (1994). *Digital Image Processing Methods*, 1st Edition, CRC Press.

3. Bhandary, A., Prabhu, G.A., Rajinikanth, V., Thanaraj, K.P., Satapathy, S.C., Robbins, D.E., Shasky, C., Zhang, Y.D., Tavares, J.M.R.S. & Raja, N.S.M. (2020). Deep-learning framework to detect lung abnormality–A study with chest X-Ray and lung CT scan images. *Pattern Recognition Letters*, 129, 271–278.

4. Dey, N., Fuqian Shi, F. & Rajinikanth, V. (2019). Leukocyte nuclei segmentation using entropy function and Chan-Vese approach. *Information Technology and Intelligent Transportation Systems*, 314, 255–264.

5. Rajinikanth, V., Dey, N., Kavallieratou, E. & Lin, H. (2020) Firefly algorithm-based Kapur's thresholding and Hough transform to extract leukocyte section from hema-tological images. In: Dey N. (eds) *Applications of Firefly Algorithm and Its Variants. Springer Tracts in Nature-Inspired Computing*, 221–235. Springer, Singapore.

6. Rezatofighi, S.H. & Soltanian-Zadeh, H. (2011). Automatic recognition of five types of white blood cells in peripheral blood. *Computerized Medical Imaging and Graphics*, 35(4), 333–343.

7. LISC. http://users.cecs.anu.edu.au/~hrezatofighi/Data/Leukocyte%20Data.htm (Accessed on: 15 March 2020).

8. Balan, N.S., Kumar, A.S., Raja, N.S.M. & Rajinikanth, V. (2016). Optimal multilevel image thresholding to improve the visibility of Plasmodium sp. in blood smear images. *Advances in Intelligent Systems and Computing*, 397, 563–571.

9. Manickavasagam, K., Sutha, S. & Kamalanand, K. (2014). An automated system based on 2d empirical mode decomposition and k-means clustering for classification of Plasmodium species in thin blood smear images. *BMC Infectious Diseases*, 14(Suppl 3), P13. doi: 10.1186/1471-2334-14-S3-P13.

10. www.dpd.cdc.gov/dpdx/HTML/ImageLibrary/Malaria_il.htm. (Accessed on: 15 March 2020).

11. Lakshmi, V.S., Tebby, S.G., Shriranjani, D. & Rajinikanth, V. (2016). Chaotic cuckoo search and Kapur/Tsallis approach in segmentation of *T. cruzi* from blood smear images. *International Journal of Computer Science and Information Security (IJCSIS)*, 14(CIC 2016), 51–56.

12. http://www.cdc.gov/dpdx/trypanosomiasisAmerican/gallery.html. (Accessed on: 15 March 2020).

13. https://uwaterloo.ca/vision-image-processing-lab/research-demos/skin-cancer-detection. (Accessed on: 15 March 2020).

14. Jesline, Rose, A.J.T., Francelin, V.S. & Rajinikanth, V. (2020). Development of a semiautomated evaluation procedure for dermoscopy pictures with hair segment. *Advances in Intelligent Systems and Computing*, 1119, 283–292.

15. Rajinikanth, V., Raja, N.S.M. & Arunmozhi, S. (2019). ABCD rule implementation for the skin melanoma assesment–A study. In: *IEEE International Conference on System, Computation, Automation and Networking (ICSCAN)*, 1–4.

16. https://polyp.grand-challenge.org/CVCClinicDB/. (Accessed on: 15 March 2020).

17. Dey, N., Shi, F. & Rajinikanth, V. (2020). Image examination system to detect gastric polyps from endoscopy images. *Information Technology and Intelligent Transportation Systems*, 323, 107–116.

18. Fernandes, S.L., Rajinikanth, V. & Kadry, S. (2019). A hybrid framework to evaluate breast abnormality using infrared thermal images. *IEEE Consumer Electronics Magazine*, 8(5), 31–36.

19. Raja, N.S.M., Rajinikanth, V., Fernandes, S.L. & Satapathy, S.C. (2017). Segmentation of breast thermal images using Kapur's entropy and hidden Markov random field. *Journal of Medical Imaging and Health Informatics*, 7(8), 1825–1829.

20. Meyer, C.R., Chenevert, T.L., Galbán, C.J., Johnson, T.D., Hamstra, D.A., Rehemtulla, A. & Ross, B.D. (2015). Data From RIDER_Breast_MRI. *The Cancer Imaging Archive*. doi: 10.7937/K9/TCIA.2015.H1SXNUXL.

21. Clark, K., Vendt, B., Smith, K., et al. (2013). The Cancer Imaging Archive (TCIA): Maintaining and operating a public information repository, *Journal of Digital Imaging*, 26(6), 1045–1057.
22. http://peipa.essex.ac.uk/info/mias.html. (Accessed on: 15 March 2020).
23. Nair, M.V., Gnanaprakasam, C.N., Rakshana, R., Keerthana, N. & V Rajinikanth, V. (2018). *International Conference on Recent Trends in Advance Computing (ICRTAC), IEEE*, 174–179.
24. http://www.onlinemedicalimages.com/index.php/en/. (Accessed on: 15 March 2020).
25. Thanaraj, R.I.R., Anand, B., Rahul, A.J. & Rajinikanth, V. (2020). Appraisal of breast ultrasound image using Shannon's thresholding and level-set segmentation. *Advances in Intelligent Systems and Computing*, 1119, 621–630.
26. http://visual.ic.uff.br/en/proeng/thiagoelias/#. (Accessed on: 15 March 2020).
27. Silva, L.F. Saade, D.C.M., Sequeiros, G.O., & Silva, A. (2014). A new database for breast research with infrared image. *Journal of Medical Imaging and Health Informatics*, 4(1), 92–100.
28. Rajinikanth, V., et al. (2018). Thermogram assisted detection and analysis of ductal carcinoma in situ (DCIS). In: *International Conference on Intelligent Computing, Instrumentation and Control Technologies (ICICICT), IEEE*, 1641–1646.
29. Aksac, A., Demetrick, D.J., Ozyer, T., et al. (2019). BreCaHAD: A dataset for breast cancer histopathological annotation and diagnosis. *BMC Research Notes*, 12, 82.
30. Rajinikanth, V., Raj, A.N.J., Thanaraj, K.P. & Naik, G.R. (2020). A customized VGG19 network with concatenation of deep and handcrafted features for brain tumor detection. *Applied Sciences*, 10(10), 3429.
31. Pugalenthi, R., Rajakumar, M.P., Ramya, J. & Rajinikanth, V. (2019). Evaluation and classification of the brain tumor MRI using machine learning technique. *Journal of Control Engineering and Applied Informatics*, 21, 12–21.
32. Rajinikanth, V., Satapathy, S.C., Fernandes, S.L. & Nachiappan, S. (2017). Entropy based segmentation of tumor from brain MR images—A study with teaching learning based optimization. *Pattern Recognition Letters*, 94, 87–95.
33. Rajinikanth, V., Raja, N.S.M. & Kamalanand, K. (2017). Firefly algorithm assisted segmentation of tumor from brain MRI using Tsallis function and Markov random field. *Journal of Control Engineering and Applied Informatics*, 19, 97–106.
34. https://www.upf.edu/web/mdm-dtic/-/1st-test-dataset#.Xupje0Uza6k (Accessed on: 15 March 2020).
35. Jahmunah, V., et al. (2019). Automated detection of schizophrenia using nonlinear signal processing methods. *Artificial Intelligence in Medicine*, 100, 101698.
36. Wang, Z. & Oates, T. (2015). Imaging time-series to improve classification and imputation. Cornell University, ArXiv:1506.00327.
37. Pedano, N., Flanders, A.E., Scarpace, L., Mikkelsen, T., Eschbacher, J.M., Hermes, B. & Ostrom, Q. (2016). Radiology data from the cancer genome atlas low grade glioma [TCGA-LGG] collection. *Cancer Imaging Archive*.
38. http://www.itksnap.org/pmwiki/pmwiki.php (Accessed on: 15 March 2020).
39. ISLES. (2015). www.isles-challenge.org. (Accessed on: 15 March 2020).
40. Maier, O., et al. (2017). ISLES 2015—A public evaluation benchmark for ischemic stroke lesion segmentation from multispectral MRI. *Medical Image Analysis* 35, 250–269.
41. Maier, O., Schröder, C., Forkert, N.D., Martinetz, T. & Handels, H. (2015). Classifiers for ischemic stroke lesion segmentation: a comparison study. *PLoS One* 10(12), e0145118.
42. Rajinikanth, V., Satapathy, S.C., Dey, N. & Vijayarajan, R. (2018). DWT-PCA image fusion technique to improve segmentation accuracy in brain tumor analysis *Lecture Notes in Electrical Engineering*, 471, 453–462.
43. Rajinikanth, V., Thanaraj, K.P., Satapathy, S.C., Fernandes, S.L. & Dey, N. (2019).

Shannon's entropy and watershed algorithm-based technique to inspect ischemic stroke wound. *Smart Innovation, Systems and Technologies*, 105, 23–31.

44. Case courtesy of Dr Derek Smith, Radiopaedia.org, rID: 36667 (Accessed on: 15 March 2020).

45. Case courtesy of Dr Henry Knipe, Radiopaedia.org, rID: 38707 (Accessed on: 15 March 2020).

46. Priya, E. & Srinivasan, S. (2018). Automated method of analysing sputum smear tuberculosis images using multifractal approach. In: Kolekar, M.H. & Kumar, V. (eds) *Biomedical Signal and Image Processing in Patient Care*, ch. 10, 184–215, IGI Global.

47. https://hc18.grand-challenge.org/.

48. Priya, E., Srinivasan, S. & Ramakrishnan, S. (2012). Differentiation of digital TB images using texture analysis and RBF classifier. *Biomedical Sciences Instrumentation*, 48, 516–523.

49. Shree, T.D.V., Revanth, K., Raja, N.S.M. & Rajinikanth, V. (2018). A hybrid image processing approach to examine abnormality in retinal optic disc. *Procedia Computer Science*, 125, 157–164.

50. Irvin, J., et al. (2019). CheXpert: A large chest radiograph dataset with uncertainty labels and expert comparison. arXiv:1901.07031 [cs.CV].

51. Wang, X., Peng, Y., Lu, L., Lu, Z., Bagheri, M., & Summers, R.M. (2017). Chest x-ray: Hospital-scale chest x-ray database and benchmarks on weakly-supervised classification and localization of common thorax diseases. arXiv: 1705.02315.

52. Case courtesy of Dr Hani Salam, Radiopaedia.org, rID: 12437 (Accessed on: 15 March 2020).

53. Case courtesy of Dr Mark Holland, Radiopaedia.org, rID: 20025 (Accessed on: 15 March 2020).

54. Case courtesy of Dr Evangelos Skondras, Radiopaedia.org, rID: 20007 (Accessed on: 15 March 2020).

55. Case courtesy of Dr Bruno Di Muzio, Radiopaedia.org, rID: 16033 (Accessed on: 15 March 2020).

56. Case courtesy of Dr David Cuete, Radiopaedia.org, rID: 27924 (Accessed on: 15 March 2020).

57. Case courtesy of Dr David Cuete, Radiopaedia.org, rID: 33983 (Accessed on: 15 March 2020).

58. https://wiki.cancerimagingarchive.net/display/Public/LIDC-IDRI (Accessed on: 15 March 2020).

59. http://segchd.csail.mit.edu/ (Accessed on: 15 March 2020).

60. Lin, H. & Rajinikanth, V. (2019). Development of softcomputing tool to evaluate heart MRI slices. *International Journal of Computer Theory and Engineering*, 11(5), 80–83.

61. Case courtesy of Assoc Prof Frank Gaillard, Radiopaedia.org, rID: 4947 (Accessed on: 15 March 2020).

62. Case courtesy of Dr Edgar Lorente, Radiopaedia.org, rID: 75187.

63. Case courtesy of Dr Edgar Lorente, Radiopaedia.org, rID: 75188.

64. Case courtesy of Dr Derek Smith, Radiopaedia.org, rID: 75249.

65. Bakiya, A. & Kamalanand, K. (2020). Mathematical modelling to assess the impact of lockdown on COVID-19 transmission in India: Model development and validation. *JMIR Public Health and Surveillance*, 6(2), e19368.

66. Wang, C., Horby, P.W., Hayden, F.G. & Gao, G.F. (2020). A novel coronavirus outbreak of global health concern. *Lancet*, 395, 470–473.

67. Nascimento, I.B.D. (2020). Novel coronavirus infection (COVID-19) in humans: A scoping review and meta-analysis. *Journal of Clinical Medicine*, 9(4), 941.

68. Fang, Y., Zhang, H., Xu, Y., Xie, J., Pang, P. & Ji, W. (2020). CT manifestations of two cases of 2019 novel coronavirus (2019-nCoV) pneumonia. *Radiology*, 295, 208–209.

69. Dey, N., Rajinikanth, V., Fong, S.J., Kaiser, M.S. & Mahmud, M. (2020). *Social-Group-Optimization Assisted Kapur's Entropy and Morphological Segmentation for Automated Detection of COVID-19 Infection from Computed Tomography Images. Preprints* 2020050052.

70. Rajinikanth, V., Dey, N., Raj, A.N.J., Hassanien, A.E., Santosh, K.C. & Raja, N.S.M. (2020). Harmony-search and otsu based system for coronavirus disease (COVID-19) detection using lung CT scan images. *arXiv preprint*, arXiv:2004.03431.

2 Image Examination

The universal disease diagnostic process followed in hospitals involves (i) personal assessment by a skilled physician and (ii) recording and assessment of illness and its symptoms using a specific protocol. In most cases, the imaging procedures are extensively considered to assess the illness in organs by means of a suitable or preferred image modality [1–3].

The preliminary level, which intends to reveal the body function with imaging, is known as the raw or unprocessed image. These images are contaminated with artifacts, along with the normal segment to be examined. To eliminate the artifact and to enhance the image segment, it is important to implement an image pre-processing technique which improves the condition of the image. This section discusses primarily used image enrichment techniques to pre-process the raw recorded image.

2.1 CLINICAL IMAGE ENHANCEMENT TECHNIQUES

Once the image is recorded using a preferred protocol, it is treated by procedures such as image enhancement, pre-processing, post-processing, and examination. This section presents the widely adopted image enhancement procedures presented in the literature. These pre-processing techniques enhance the medical images recorded to aid in the examination of abnormalities.

Usually, an image of a preferred dimension and a selected modality provides details in the form of visual representation. According to the image registration, the picture is classified as two-dimensional (2D) or three-dimension (3D). Normally, the processing actions existing for 2D imagery are moderately straightforward compared to 3D. Moreover, these images are additionally classified as usual (recorded with gray/RGB scale pixels) or binary images.

In various fields, images recorded with a selected modality through preferred pixel significance can be used to allocate significant information. In some circumstances, the information existing in the untreated images are difficult to analyze. Therefore, pre-processing and post-processing techniques are proposed and executed by researchers [4,5]. The implemented picture processing methods can help develop the state of the unprocessed image with a selection of methodologies such as contrast-enhancement, edge-detection, noise-removal, filtering, fusion, thresholding, segmentation, etc. [6–8].

Most of the existing augmentation events work well for greyscale pictures compared to RGB-scale pictures. In the literature, several picture assessment measures are available to pre-process the test images, a procedure that can help translate the raw test image into a satisfactory test image [9,10]. The need for image augmentation and its practical implication is clearly discussed in the subsequent sections for both the greyscale and the RGB-scale pictures.

2.2 IMPORTANCE OF IMAGE ENHANCEMENT

Initially, images recorded using a chosen image modality is referred to as the unprocessed image. Based on what is required, these raw images may be treated with a specific image conversion or enrichment technique. The digital images recorded using the well-known imaging methodologies are associated with various problems. Before further assessment, it is essential to improve information available in the image [11–13]. Presently, recorded digital images are processed and stored using digital electronic devices. To ensure eminence, a particular image modification procedure is needed to convert the raw image into the processed image. Enhancement procedures, such as (i) artifact removal, (ii) filtering, (iii) contrast enrichment, (iv) edge detection, (v) thresholding, and (vi) smoothing, are some common procedures implemented in the literature to convert the unprocessed image. Image enhancement is essential to improve the visibility of recorded information. Also, extracting recorded information from enhanced images is easier compared with information from unprocessed images.

2.3 INTRODUCTION TO ENHANCEMENT TECHNIQUES

Most recent medical imaging systems are computer-controlled systems. Thus, the images attained by means of imaging schemes are digital. The excellence of an image is judged based on the visibility of Section-of-Interest (SOI) and the distinct variation between the background and the SOI. The picture recorded by a particular imaging mechanism needs to be processed to convert it into a working image. This procedure is essential and a complex task when the image SOI is associated with unwanted noise and artifact [14–16].

A number of image enhancement techniques are proposed and implemented in the literature by the researchers, some of which are explained below.

2.3.1 ARTIFACT REMOVAL

Artifact removal is essential to divide the picture into a number of sub-sections based on an optimal threshold value. The artifact elimination technique usually implements a morphological filter along with a clustering approach such that the image pixels are ordered, grouped to extract a variety of information based on the selected threshold value. This approach is widely implemented in medical image processing applications to remove artifacts in the 2D slices of Magnetic-Resonance-Image (MRI) and Computed-Tomography (CT). In this section, the 2D brain MRI slice of the MRI and the CT scan slice are considered to demonstrate the performance of the selected threshold filter.

Figure 2.1 presents the structure of the procedure. Figure 2.2 (a) and (b) illustrates the results attained by the brain MRI and the lung CT scan images, respectively.

Threshold is a type of filter used to pre-process the medical images recorded with MRI or CT. Several previous works on the brain MRI with threshold-filter approach have been reported. This separates the brain MRI slices into skull section and the normal brain anatomical regions. After pre-processing, the test image with the recommended technique is further considered for the examination process. Literature confirms that the pre-processed medical images (MRI as well as CT) with the filter being implemented help attain better results compared to the unprocessed image [2,17].

Advantages: Threshold filter separates ROI and artifact and it diminishes the difficulty in assessing the raw image.

Limitations: The major restriction of the threshold filter is the choice of optimal threshold, which segregates the raw image into two sections. In most cases, the threshold choice is done physically with a range of trials. Experiment-based method is a time-consuming practice which works only on greyscale images.

2.3.2 Noise Removal

In conventional digital image processing, the filter employed through a selected practice and a favorite order is to allow/block the image information based on the frequency value. In this operation, unnecessary pixels accessible in the digital picture is removed/blocked with a preferred filter based on pixel operation [18,19].

Figure 2.3 presents the arrangement of the conventional filter, which eradicates noise from the digital gray/RGB image. Figure 2.4 presents the results attained for a 2D brain MRI slice. In this work, the noise filter is employed to eliminate the noise (Salt & Pepper) associated with the test image. Once the noise is eliminated, the filtered image becomes clear for further processing. Results obtained in various image modalities are depicted in Figure 2.5 which may be gray or RGB in nature.

Advantages: The image filter removes surplus pixels in the chosen digital representation. It can also be used as an essential pre-processing system for the medical images of varied modalities.

Limitations: Attainment of a selective image filter to assess the image is relatively time-consuming as it processes the raw images stained with noise.

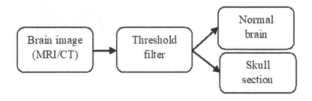

FIGURE 2.1   Implementation of Threshold-Filter for ROI and Artifact Separation.

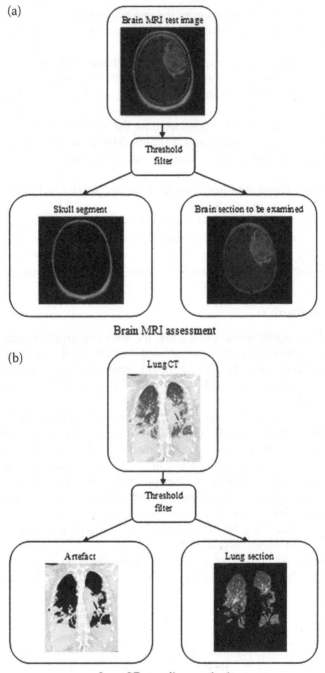

FIGURE 2.2   Experimental Results of the Threshold Filter.

FIGURE 2.3   Implementation of a Filter to Eliminate the Associated Noise.

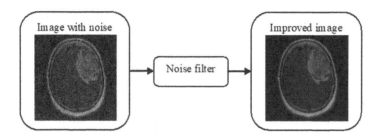

FIGURE 2.4   Salt and Pepper Noise Removal Implemented on a Brain MRI Slice.

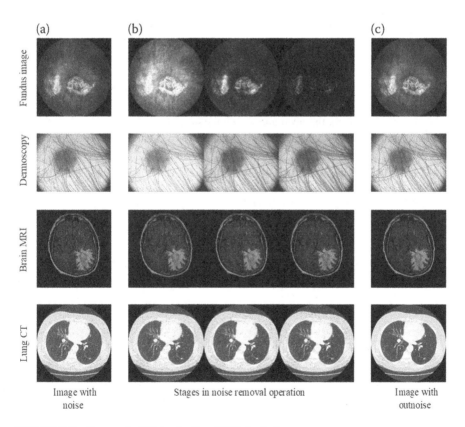

FIGURE 2.5   Removal of Noise in Gray/RGB Scale Image.

2.3.3 CONTRAST ENRICHMENT

Information present in the greyscale image is usually poor compared with that of the RGB-scale image. In greyscale images, enhancing the ROI with respect to its surroundings is a critical task, thus a number of Image Processing Schemes (IPS) are proposed and implemented by the researchers to improve visibility. Contrast enhancement is one of the widely adopted techniques in medical image processing. This can be implemented in a number of ways. The schemes, such as histogram equalization, Color-map tuning and Contrast-Limited-Adaptive-Histogram-Equalization (CLAHE), are some of the procedures considered to enhance greyscale images [14,18].

Figures 2.6 and 2.7 depict the structure of the IPS and its related results, respectively. It is observed from these figures that the results of the considered procedures help enhance the visibility of SOI compared to the unprocessed image. The experimental study with the brain MRI slice confirms that the considered approaches improve the visibility of the tumor segment substantially. Later, the improved SOI can be extracted and evaluated using an appropriate segmentation procedure.

The image enhancement technique implemented on the raw test image will enhance the SOI section by adjusting its pixel distribution (adjusting the histogram) based on the implemented technique. The various shapes of the gray-level histogram of the brain MRI for various techniques are depicted in Figure 2.8 and the comparison of Figure 2.8 (a) with Figure 2.8 (b) to (d) depicts that variation in the threshold will enhance the SOI by modifying pixel distribution. The choice of pixel adjustment depends on the need and the expertise of the operator who executed the enhancement process. Contrast enhancement employed for other gray/RGB-scale test images with and without noise is projected in Figure 2.9. The augmentation procedure discussed in this section is suitable for any gray/RGB class image with or without noise.

Advantages: Image contrast enhancement is a common technique and requires low computational effort during implementation.

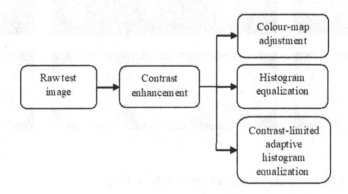

FIGURE 2.6   Commonly Considered Contrast Enhancement Methods.

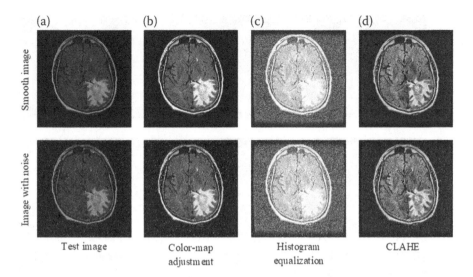

FIGURE 2.7   Contrast Enhancement of the Brain MRI Slice to Enhance the Tumor Section.

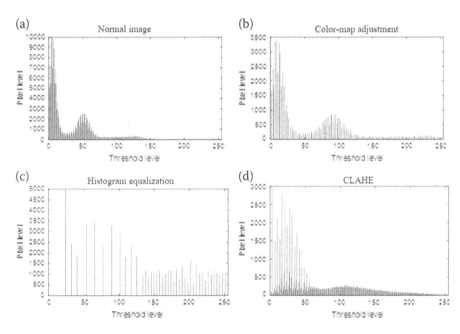

FIGURE 2.8   Variation of Pixel Distribution in Gray-Level Threshold Based on the Implemented Contrast Enhancement Technique.

Limitations: This technique is used throughout the preliminary-level picture augmentation process and in most of image processing. It is an optional practice and the images connected with noise will not offer expected results with this procedure.

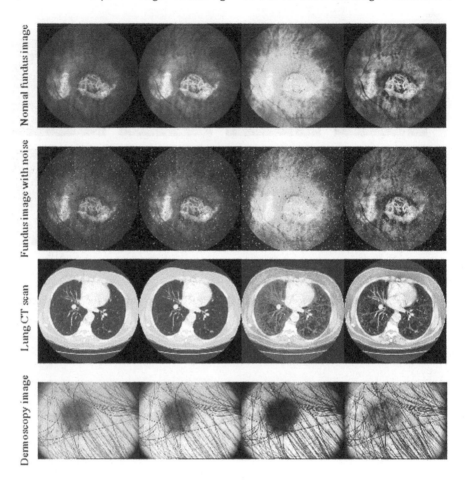

FIGURE 2.9   Contrast Enhancement Technique Implemented on a Class of Images.

2.3.4 Edge Detection

This process is used to outline the borders of the SOI existing in the test picture. In the literature, a number of edge detection methods have been listed. The Canny-Edge-Detector (CED) is generally a popular method starting in 1986 [15]. This employs a multi-stage algorithm to recognize a broad collection of edges in the images. CED is used to extract precious structural information from the images but considerably diminishes the quantity of information to be investigated. The general circumstance for edge recognition includes:

a.   Recognition of a border with smaller error rate by accurately finding numerous edges existing in the trial image.
b.   The perimeter tip predictable from the operator should accurately be confined to the heart of the border.

c.   The perimeter of the picture should only be recognized once, and the image noise should not form fake boundaries.

To achieve these points, Canny implemented calculus of variations to optimize the operation.

Sobel is another common type of edge detection procedure adopted in the image processing literature [20–22]. The edge detection result attained with Sobel and Canny on a lung CT image is presented in Figure 2.10. It is noticed from Figure 2.10 (b) that the investigational result of the Canny is superior to Sobel and the selection of particular edge discovery can be chosen as per the expertise of the operator.

Advantage: Implementation of edge discovery is necessary to identify the boundary and the texture of the picture under study.

Limitation: The edge detection process necessitates complex procedures to recognize the boundary of the picture and this practice will not offer fitting outcomes when RGB-scale images are considered.

2.3.5 RESTORATION

Haziness in medical images arises due to various reasons . This problem degrades the information in the images. This is a common problem that can be overcome by performing the imaging procedure once again. During the crucial image recording process, retaking of the image is not possible; further, image recording requires a considerable number of procedures and precautionary measures. Hence, it is necessary to implement image restoration techniques which help to correct/remove the haziness in the medical image. The restoration technique will compute the major pixels in the image and fix the distortion to get a better restoration. The sample test images considered for demonstration is depicted in Figure 2.11 and the corresponding results attained with MATLAB supported restoration process is depicted in Figure 2.12. Figure 2.12 (d) confirms that the considered restoration technique helped to correct the test image of various modalities.

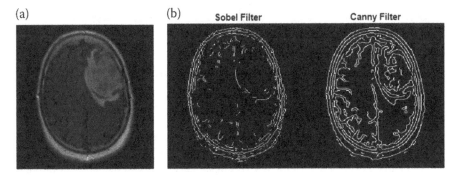

FIGURE 2.10   Result Attained with the Edge Detection Technique. (a) Test Image, and (b) Attained Result with Sobel and Canny Filters.

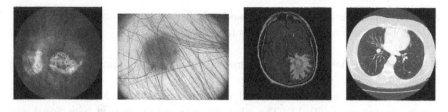

FIGURE 2.11 Sample Test Images for the Assessment.

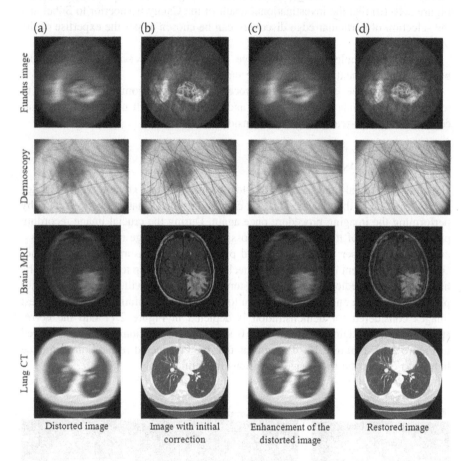

FIGURE 2.12 Result Attained with the Edge Detection Technique.

2.3.6 COLOR SPACE CORRECTION

Color space correction is an essential procedure implemented to regulate the R, G, and B pixels in RGB-scale pictures to improve the SOI of the test picture. In this work, the pixel distribution of the R/G/B thresholds is adjusted manually or through a computer algorithm to enhance the results. The chosen test image and the results obtained after the color space correction operation is depicted in Figure 2.13.

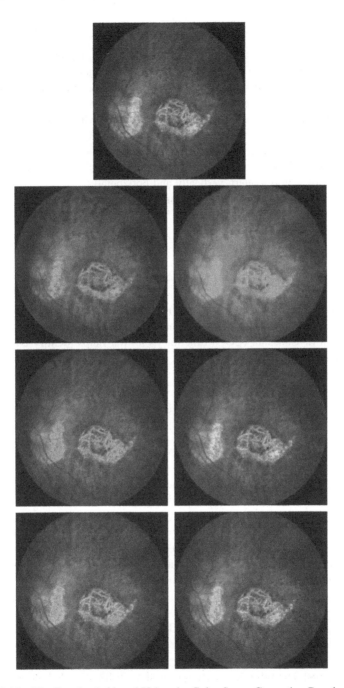

FIGURE 2.13 The Results Achieved Using the Color Space Correction Practice.

2.3.7 IMAGE EDGE SMOOTHING

During automatic recognition and classification operations, texture and silhouette features extracted from the picture play a fundamental function. Before extracting the surface features from a picture, it is necessary to treat the raw image with preferred picture normalization procedures. The image's surface smoothing based on a chosen filter is widely adopted to improve the texture features. Earlier works confirmed that the Gaussian-Filter (GF) practice can be employed to recover the surface and the perimeter features of the greyscale picture for a selected dimension [18]. Further, one of the previous research confirmed that the GF with varied scale (ϕ) can provide enlargement of texture pattern vertically and horizontally.

The conventional Gaussian operator for a 2D image is presented as in Equation (2.1):

$$U(x,\ y) = \frac{1}{2\pi\phi^2} e^{-\left[\frac{x^2+y^2}{2\phi^2}\right]} \tag{2.1}$$

where ϕ = standard deviation and $U(x, y)$ = Cartesian coordinates of the image. By altering ϕ, imagery with diverse edge enhancements can be created.

Laplacian of Gaussian (LOG) function as a filter helps identify edges by computing the zero-crossings of their second derivatives as observed in Equation (2.2).

$$\nabla^2 U(x,\ y) = \frac{d^2}{dx^2}U(x,\ y) + \frac{d^2}{dy^2}U(x,\ y) = \frac{x^2 + y^2 - 2\phi^2}{2\pi\phi^6} e^{-\left[\frac{x^2+y^2}{2\phi^2}\right]} \tag{2.2}$$

Figure 2.14 presents the selected brain MRI and the related experimental results. Figure 2.14 (a) presents the horizontally smoothened MRI slice for a chosen value of $\phi = 30$ and Figure 2.14 (b) depicts the vertically smoothened image where $\phi = 300$. After augmentation, the surface and edge features from these images are extracted. These features are then considered to train, test, and validate the classifier arrangement employed to detect/classify the tumor seen in the brain MRI. The images treated with the GF shows a variety of texture and edge values for both the SOI. The other sections of the picture and assessment of these values present a better understanding of the images and its irregularity.

Advantages: GF-based practices are used to regularize the surface and perimeter of the greyscale images of varied aspects. Further, this scheme can be used to create various edge as well as texture patterns based on the chosen ϕ. The GF also supports a wide range of edge and texture detection procedures. Canny edge discovery practice employs GF to recover the edges of the test image.

Limitations: The information present in the GF filter treated test image can be examined only with a selected image examination practice as the information cannot be examined by physical operators. Hence, this image assessment can be used only when a computer-assisted algorithm is employed to detect/classify the picture based on its SOI's edge/texture features.

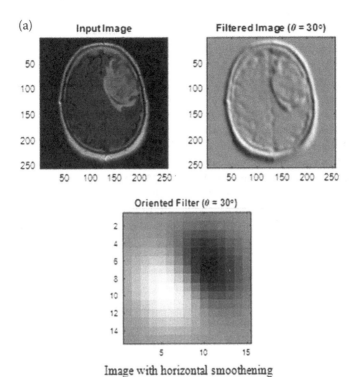

Image with horizontal smoothening

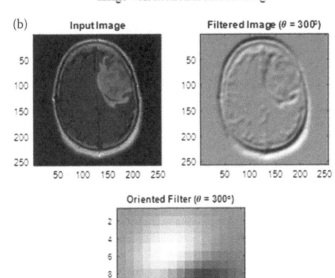

Image with vertical smoothening

FIGURE 2.14   Results Attained with Gaussian-Filter-Based Enhancement.

2.4 RECENT ADVANCEMENTS

Presently, image examination is executed by an experienced worker and the attained result requires some negotiation. Due to the rapid progression in modern technology, the use of computers has become frequent in almost all domains due to its simplicity and adaptability. Recently, a considerable number of computer algorithms have been developed to support a variety of image examination operations. The computerized image processing (i) supports semi-automated/automated examination, (ii) works well for greyscale/RGB images, (iii) can be used to implement a variety of soft-computing techniques, and (iv) the results of these methods can be stored temporarily or permanently based on the requirement [23–28].

The computerized algorithms helped achieve a variety of image assessment procedures to improve the clarity and the information in the picture irrespective of its color, size, and pixel distribution. Further, a number of software have been developed to maintain the computerized image processing procedures without compromising its excellence and the throughput with these procedures are better compared to operator-assisted schemes.

Due to its qualities, computer-assisted procedures are employed in a variety of image-based assessment schemes to perform necessary operations such as artifact removal, contrast enhancement, edge detection, smoothing, and thresholding. Further, the availability of heuristic algorithms and its support towards image processing has helped to expand and implement different machine-learning and deep-learning techniques to process a selection of images with a preferred pixel dimension.

The remaining part of this section presents information on the hybrid image processing implementation to assess the disease from a class of clinical-grade medical images.

2.4.1 HYBRID IMAGE EXAMINATION TECHNIQUE

Recent image processing literature confirms the need for hybrid image examination techniques to achieve better disease detection using the test images of a chosen modality.

The hybrid image examination technique combines all the possible image processing procedures to improve detection accuracy. The common structure of the hybrid-image examination procedure implemented with a chosen heuristic algorithm is depicted in Figure 2.15. The main task of this structure is to implement a particular image enhancement technique to improve the visibility of the SOI. Then, a chosen segmentation procedure is implemented to extract the infected section from the test image. Finally, the extracted information is verified by a doctor, who then implements treatment based on the severity.

2.4.2 NEED FOR MULTI-LEVEL THRESHOLDING

Image thresholding based on a selected guiding function is extensively implemented in a variety of fields to pre-process the examination image. An image could be known as the arrangement of different pixels with reference to the thresholds. In a digital

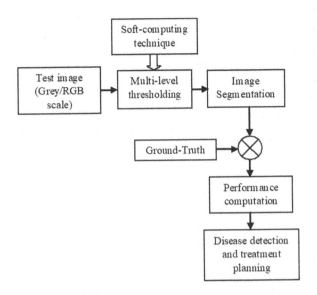

FIGURE 2.15   Structure of the Hybrid-Image Processing Methodology.

image, the pixel allotment plays a chief role and adjusting or grouping these pixels are preferred to enhance/change the information existing in the illustration.

Earlier, bi-level thresholding is used to separate the raw picture into SOI and the background. In this process, the operator/computer algorithm is allowed to recognize a Finest-Threshold (FT) by means of a preferred function. Let the specified image have 256 thresholds (ranging from 0 to 255) and from this, one threshold value is selected as the Finest-Threshold as follows;

$$0 < FT < 255 \qquad (2.3)$$

The selected threshold will assist in dividing the image into two sections, such as section 1 with pixel distribution <FT and section 2 with pixel distribution >FT. This procedure divides the image into two-pixel groups and, therefore, is called bi-level thresholding.

If the plan is to divide the specified image into more than two sections, then a multi-level thresholding is chosen. During this operation, the number of FT = number of threshold levels. For example, if a tri-level threshold is chosen for the study, it will separate the test image into three sections as depicted in Equation (2.4)

$$0 < FT_1 < FT_2 < 255 \qquad (2.4)$$

In this case, the image is separated into three sections, such that section1 (pixels between 0 to FT_1), section 2 (pixels between FT_1 to FT_2), and section 3 (pixels between FT_2 to 255). In most applications, the information attained with the bi-level threshold is not appropriate. Hence, multi-thresholding is preferred to pre-process greyscale/RGB images. The information on multi-level threshold can be found in the literature [29–32].

2.4.3 THRESHOLDING

Image SOI enhancement based on the threshold value is extensively adopted in the literatures to practice a class of traditional and medical imagery. In most of the illustration assessment systems, thresholding is adopted as the pre-processing approach. In the thresholding procedure, the SOI in a chosen image (greyscale/RGB) is improved by grouping the pixels based on the selected optimal threshold value. The grouping of the image pixel will divide the SOI from the image background and other parts, and after this enhancement, the SOI can be extracted using a segmentation process. Figure 2.16 depicts the bi-level and multi-level threshold attained for a chosen test image with varied modalities. Figure 2.16 (a) and (b) presents the threshold results attained for a greyscale and RGB-scale image.

The literature based confirms the availability of the thresholding procedure based on traditional and soft-computing driven techniques. In the traditional/operator-supported technique, the optimum threshold recognition is done manually through trials. During this process, the threshold of the test image is rapidly varied with its histogram and different pixel grouping is studied to achieve the enhanced illustration. Identification of the optimum threshold through predictable process is complex and time-consuming. To overcome this difficulty, recently, threshold operation is performed under the supervision of soft-computing algorithms. In hybrid medical data evaluation techniques, thresholding is adopted as the pre-processing method to improve the SOI [33–35].

Based on the selected number of thresholds, it is classified as bi-level (separating the image into SOI and background) and multi-level threshold (dividing the into various pixel groups). The threshold result obtained on a specific brain MRI and fundus image is depicted in Figure 2.16. The result outcome confirms the thresholding process that enhances the visibility of the infected section.

Advantages: Thresholding is a straightforward and efficient image enhancement procedure and it can be achieved both manually and with the help of heuristic algorithms. Further, it supports the processing of gray/RGB-scale images with varied schemes. SOI of the test image can be improved by either a bi-level or a multi-level threshold selection process.

Limitations: The choice of the optimum threshold and the objective function throughout the soft-computing assisted thresholding is fairly complicated. Further, in most image enhancement techniques, the thresholding is preferred only as the pre-processing technique. The computation time of this operation increases based on the histogram complexity and the pixel distribution.

2.4.4 IMPLEMENTATION AND EVALUATION OF THRESHOLDING PROCESS

Optimal threshold-based test image enhancement can be executed using manually or with a computer algorithm. The computation complexity in operator-assisted FT selection process needs more effort. Further, this complexity increases if the number of FT to be identified is >2 (i.e. multi-threshold selection). The operator adjusts the thresholds arbitrarily until the best image separation is achieved. Thus, in recent years, heuristic algorithm-assisted image thresholding has been widely

(a)   (b)

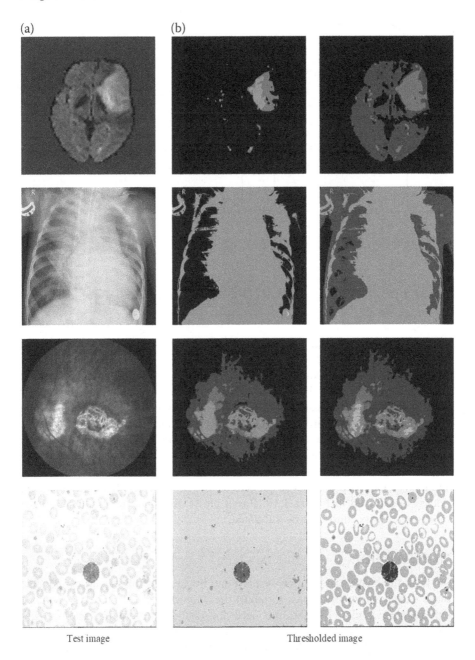

Test image   Thresholded image

FIGURE 2.16 Multi-Level Threshold Implemented on Gray/RGB-Scale Image.

proposed by researchers. In this method, a chosen heuristic-algorithm with a chosen objective function is implemented to enhance image information.

After implementing thresholding using a chosen procedure, its outcome needs to be validated to verify the eminence of the proposed procedure. This validation

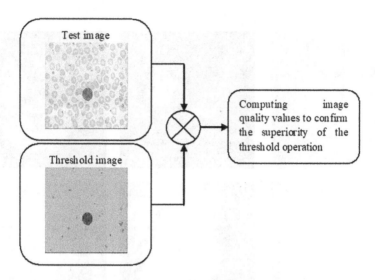

FIGURE 2.17   Performance Validation of the Thresholding Process.

can be achieved through a comparative analysis between the raw image and the threshold image. During this process, commonly considered image quality values such as root-mean-squared-error (RMSE), peak-signal-to-noise-ratio (PSNR, in dB), structural-similarity-index (SSIM), normalized-absolute-error (NAE), normalized-cross-correlation (NCC), average-difference (AD), and structural-content (SC) are computed for every image. Based on these values, the performance of the implemented thresholding technique is measured. Further, the results attained with a chosen technique are compared with the results of state-of-the-art techniques existing in the literature [36,37]. The commonly employed evaluation procedure to verify the result of the threshold process is depicted in Figure 1.10. Based on its outcome, the eminence of the implemented procedure can be validated.

2.5 SUMMARY

The need for image examination systems and implementation of various possible image enhancement procedures to adjust the unprocessed test picture is outlined with suitable results. Further, the need for picture thresholding and the implementation of bi-level and multi-level thresholding are clearly discussed. Implementation of thresholding process for the gray/RGB-scale images and its assessment procedures are presented. From the discussion, it can be noted that image thresholding performed by means of a chosen method can be implemented to pre-process the digital test images recorded from a particular imaging modality. Moreover, image thresholding has emerged as a key research field due to its practical significance and has become a chief procedure to be considered in the hybrid image examination system.

REFERENCES

1. Priya, E. & Srinivasan, S. (2015). Automated identification of tuberculosis objects in digital images using neural network and neuro fuzzy inference systems. *Journal of Medical Imaging and Health Informatics*, 5, 506–512.
2. Bhandary, A., Prabhu, G.A., Rajinikanth, V., Thanaraj, K.P., Satapathy, S.C., Robbins, D.E., Shasky, C., Zhang, Y.D., Tavares, J.M.R.S. & Raja, N.S.M. (2020). *Pattern Recognition Letters*, 129, 271–278.
3. Rajinikanth, V., Dey, N., Raj, A.N.J., Hassanien, A.E., Santosh, K.C. & Raja, N.S.M. (2020). Harmony-search and otsu based system for coronavirus disease (COVID-19) detection using lung CT scan images, *arXiv preprint*, arXiv:2004.03431.
4. Priya, E., Srinivasan, S. & Ramakrishnan, S. (2014). Retrospective non-uniform illumination correction techniques in microscopic digital TB images. *Microscopy and Microanalysis*, 20(5), 1382–1391.
5. Fernandes, S.L., Rajinikanth, V. & Kadry, S. (2019). A hybrid framework to evaluate breast abnormality using infrared thermal images. *IEEE Consumer Electronics Magazine*, 8(5), 31–36.
6. Priya, E. & Srinivasan, S. (2013). Automated decision support system for tuberculosis digital images using evolutionary learning machines. *European Journal for Biomedical Informatics*, 9(1), 2–7.
7. Priya, E., Srinivasan, S. & Ramakrishnan, S. (2012). Classification of tuberculosis digital images using hybrid Evolutionary Extreme Learning Machines. In: Nguyen N.T., Hoang K., Jedrzejowicz P. (eds) *Computational Collective Intelligence. Technologies and Applications. Lecture notes in Computer Science*, Springer, Berlin, Heidelberg, 7653, 268–277.
8. Thanaraj, R.I.R., Anand, B., Rahul, A.J. & Rajinikanth, V. (2020). Appraisal of breast ultrasound image using Shannon's thresholding and level-set segmentation. *Advances in Intelligent Systems and Computing*, 1119, 621–630.
9. Raja, N.S.M., Rajinikanth,V. & Latha, K. (2014). Otsu based optimal multilevel image thresholding using firefly algorithm. *Modelling and Simulation in Engineering*, 2014, 794574, 17.
10. Akay, B. (2013). A study on particle swarm optimization and artificial bee colony algorithms for multilevel thresholding, *Applied Soft Computing Journal*, 13(6), 3066–3091.
11. Priya, E. & Srinivasan, S. (2017). Analysis of tuberculosis images using differential evolutionary extreme learning machines (DE-ELM). In: Hemanth, D.J. & Estrela, V.V. (eds) *Deep Learning for Image Processing Applications, Advances in Parallel Computing*, IOS Press, 111–136.
12. Priya, E., Srinivasan, S. & Ramakrishnan, S. (2012). Differentiation of digital TB images using texture analysis and RBF classifier. *Biomedical Sciences Instrumentation*, 48, 516–523.
13. Dey, N., et al. (2019). Social-group-optimization based tumor evaluation tool for clinical brain MRI of flair/diffusion-weighted modality. *Biocybernetics and Biomedical Engineering*, 39(3), 843–856.
14. Stark, J.A. (2000). Adaptive image contrast enhancement using generalizations of histogram equalization. *IEEE Transactions on Image Processing*, 9(5), 889–896.
15. Canny, J. (1986). A computational approach to edge detection. *IEEE Transactions on Pattern Analysis and Machine Intelligence, PAMI*, 8(6), 679–698.
16. Priya, E. & Srinivasan, S. (2016). Validation of non-uniform illumination correction techniques in microscopic digital TB images using image sharpness measures. *International Journal of Infectious Diseases*, 45(S1), 406.

17. Dey, N., Rajinikanth, V., Fong, S.J., Kaiser, M.S. & Mahmud, M. (2020). Social-Group-Optimization Assisted Kapur's Entropy and Morphological Segmentation for Automated Detection of COVID-19 Infection from Computed Tomography Images. *Preprints* 2020050052.

18. Dougherty, E.R. (1994). *Digital Image Processing Methods*, 1st Edition, CRC Press.

19. Raja, N.S.M., Rajinikanth, V., Fernandes, S.L. & Satapathy, S.C. (2017). Segmentation of breast thermal images using Kapur's entropy and hidden Markov random field. *Journal of Medical Imaging and Health Informatics*, 7(8), 1825–1829.

20. Zhou, P., Ye, W. & Wang, Q. (2011). An improved canny algorithm for edge detection. *Journal of Computational Information Systems*, 7(5), 1516–1523.

21. Basu, M. (2002). Gaussian-based edge-detection methods-a survey. *IEEE Transactions on Systems, Man, and Cybernetics, Part C: Applications and Reviews*. https://doi.org/10.1109/TSMCC.2002.804448.

22. Marr, D. & Hildreth, E. (1980). Theory of edge detection. *Proceedings of the Royal Society of London. Series A, Mathematical and Physical Sciences*, B, 207, 187–217.

23. Manikantan, K., Arun, B.V. & Yaradonic, D.K.S. (2012). Optimal multilevel thresholds based on Tsallis entropy method using golden ratio particle swarm optimization for improved image segmentation. *Procedia Engineering*, 30, 364–371.

24. Rajinikanth, V., Raja, N.S.M. & Latha, K. (2014). Optimal multilevel image thresholding: An analysis with PSO and BFO algorithms. *Australian Journal of Basic and Applied Sciences*, 8(9), 443–454.

25. Satapathy, S.C., Raja, N.S.M., Rajinikanth, V., Ashour, A.S. & Dey, N. (2018). Multi-level image thresholding using Otsu and chaotic bat algorithm. *Neural Computing and Applications*, 29(12), 1285–1307.

26. Rajinikanth, V. & Couceiro, M.S. (2015). RGB histogram based color image segmentation using firefly algorithm. *Procedia Computer Science*, 46, 1449–1457.

27. Nair, M.V., Gnanaprakasam, C.N., Rakshana, R., Keerthana, N. & Rajinikanth, V. (2018). *International Conference on Recent Trends in Advance Computing (ICRTAC)*, IEEE, 174–179.

28. Rajinikanth, V., Raja, N.S.M. & Arunmozhi, S. (2019). ABCD rule implementation for the skin melanoma assesment–A study, In: *IEEE International Conference on System, Computation, Automation and Networking (ICSCAN)*, 1–4.

29. Rajinikanth, V., Thanaraj, K.P., Satapathy, S.C., Fernandes, S.L. & Dey, N. (2019). Shannon's entropy and watershed algorithm based technique to inspect ischemic stroke wound. *Smart Innovation, Systems and Technologies*, 105, 23–31.

30. Shree, T.D.V., Revanth, K., Raja, N.S.M. & Rajinikanth, V. (2018). A hybrid image processing approach to examine abnormality in retinal optic disc. *Procedia Computer Science*, 125, 157–164.

31. Dey, N., Shi, F. & Rajinikanth, V. (2020). Image examination system to detect gastric polyps from endoscopy images. *Information Technology and Intelligent Transportation Systems*, 323, 107–116.

32. Fernandes, S.L., Rajinikanth, V. & Kadry, S. (2019). A hybrid framework to evaluate breast abnormality using infrared thermal images. *IEEE Consumer Electronics Magazine*, 8(5), 31–36.

33. Ghamisi, P., Couceiro, M.S., Benediktsson, J.A. & Ferreira, N.M.F. (2012). An efficient method for segmentation of images based on fractional calculus and natural selection. *Expert Systems with Applications*, 39(16), 12407–12417.

34. Priya, E. & Srinivasan, S. (2016). Automated object and image level classification of TB images using support vector neural network classifier. *Biocybernetics and Biomedical Engineering*, 36(4), 670–678.

35. Ghamisi, P., Couceiro, M.S., Martins, F.M.L. & Benediktsson, J.A. (2014). Multilevel image segmentation based on fractional-order Darwinian particle swarm optimization. *IEEE Transactionson Geoscience and Remote sensing*, 52(5), 2382–2394.

36. Hore, A. & Ziou, D. (2010). Image quality metrics: PSNR vs. SSIM. In: *IEEE International Conference on Pattern Recognition (ICPR)*, Istanbul, Turkey, 2366–2369.

37. Wang, Z., Bovik, A.C., Sheikh, H.R. & Simoncelli, E.P. (2004). Image quality assessment: From error measurement to structural similarity. *IEEE Transactions on Image Processing*, 13(1), 1–14.

3 Image Thresholding

In the medical domain, images recorded with a preferred modality with a favored pixel value conveys significant information. In some circumstances, the information found in the unrefined images are hard to analyze, thus several pre-processing and post-processing methods are required, which are proposed and executed by the researchers [1–5]. The implemented image processing schemes improve the condition of the unprocessed image using a variety of methodologies such as edge-detection, noise-removal, contrast enrichment, and thresholding. Due to its reputation and practical significance, a variety of gray/RGB-scale image threshold selection methods are implemented by the researchers to process the digital photographs recorded through varied modalities. A preferred thresholding practice improves the visibility of a section by grouping related pixels according to the selected threshold value. The remaining sections in this chapter summarize various image thresholding procedures generally used in the literature to pre-process and enhance a range of images with varied size and pixel values.

3.1 NEED FOR THRESHOLDING OF MEDICAL IMAGES

Traditional and soft-computing-based multi-thresholding is an established image pre-processing practice extensively adopted in the image processing literature. This practice is implemented to enhance the visibility of the SOI in various test images. The threshold value can be decided for the specific test gray/RGB-scale image by plotting the histogram of the image. The histogram is a graphical demonstration of the gray values (X-axis) with respect to the pixel distribution (Y-axis). The considered threshold value throughout the image examination task is $L = 256$ for simplicity. The threshold value is fixed for every image and the pixel distribution varies depending on the size and information available in the image.

Let us consider a brain MRI slice with dimensions of $256 \times 256 \times 1$ pixels. The histogram of the image represents the distribution of the image pixels with respect to the threshold. The threshold value is attained by employing the bi-level and multi-level threshold selection process, clearly presented in Figure 3.1 for this image. In this image, Figure 3.1 (a) to (c) depicts the sample test image, histogram, and the bi-level based threshold image, respectively. It is confirmed by Figure 3.1 (c) that the thresholding procedure had improved the visibility of the tumor section. Other related information regarding the medical image thresholding can be found in research articles [6–8].

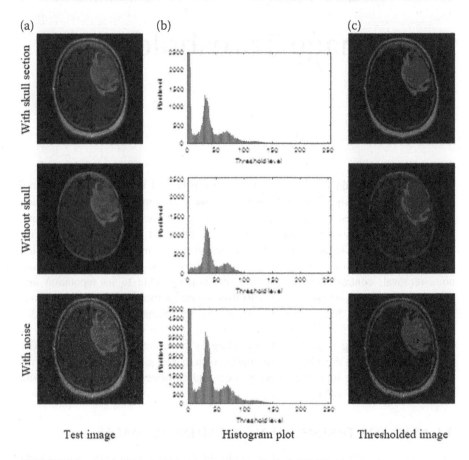

(a)   (b)   (c)

With skull section

Without skull

With noise

Test image   Histogram plot   Thresholded image

FIGURE 3.1   Image Threshold Implemented for Brain MRI Slice.

Image threshold processes help improve the pixel-level information available in the digital image, which can then be recognized and analyzed for further assessment. In medical practice, thresholding process would separate the digital image into the background, the normal section, and the section with the abnormality. Finally, the section with abnormality can be extracted with a chosen technique to ease further examination. As earlier discussed, thresholding is a well-known preprocessing technique adopted in medical image evaluation [9–13].

3.2 BI-LEVEL AND MULTI-LEVEL THRESHOLD

The aim of image thresholding is to recognize the finest/optimal threshold, which separates the picture into various classes based on the assigned threshold requirement. The threshold selection for greyscale images is straightforward compared to that of the RGB-scale. Greyscale implementation is simple, but for RGB-scale case threshold selection must be implemented for the histograms of R, G, and B separately. Depending on the number of thresholds, it is classified as (i) bi-level and (ii) multi-level threshold. This procedure is clearly discussed in upcoming sections with appropriate experimental results attained with the MATLAB software.

Because of its simplicity, the thresholding of bi-level cases is considered first as it can be extended to the multi-level with appropriate modifications. Let $T = (t_0, t_1, \ldots, t_{L-1})$ represent the number of thresholds available in a selected digital image of fixed dimensions, and each image needs to be assessed by considering its histogram created by considering its pixel (Y-axis) and threshold distribution (X-axis). The thresholding process needs the identification of an optimal threshold value $T = t_{OP}$, which supports the grouping of image pixels to enhance the visibility of ROI. The thresholding scheme can be implemented using (i) bi-level approach (sorting out the image pixels into two groups) and (ii) multi-level approach (sorting out the image pixels into multiple clusters).

- Bi-level threshold: This procedure helps separate the test image into two pixel groups, such that the ROI and the background is based on $T = t_{OP}$. Level 1 will be obtained by considering image pixels $>T = t_{OP}$ while Level 2 is obtained with image pixels $<T = t_{OP}$. To achieve this, a manual operator can be employed which can identify $T = t_{OP}$ using trial and error or the heuristic algorithm, the latter of which can solve this problem with less effort.
- Multi-level threshold: This is the extension of the bi-level operation in which the digital image is separated into more than two pixel groups. In this operation, several of t_{OP} are identified. If an image is separated with $T = t_{OP1}, \ldots, t_{OPn}$, then it is called the multi-thresholding process.

3.3 COMMON THRESHOLDING METHODS

Detection of an appropriate technique to improve the digital image is a moderately challenging task. The suitable image examination technique can be chosen based on the recommendation of previous literature or experience. In the literature, several threshold techniques are available, of which the most commonly implemented are presented in Figure 3.2. Each method has its own merits and demerits and, in the literature, histogram-assisted threshold selection procedures are mostly considered to enhance the greyscale/RGB-scale picture.

In the histogram-assisted threshold processes, the histogram is examined using a chosen technique, in which the threshold values are randomly varied until the quality of the processed image reaches optimal stage. In this technique, a selected objective value is taken as the guiding mechanism to justify the quality of the image based on a chosen image-related measure. This measure is called the objective function.

The complexity of thresholding increases due to the (i) size of the image, (ii) pixel distribution, and (iii) multiple pixel class (i.e. RGB). The number of pixels in the image also rises when the image dimension increases, which may lengthen the computation time during threshold selection. Further, uneven pixel distribution can also make thresholding more complex, such as when (arge peaks and valleys are present in the histogram. The RGB-scale histogram can also increase complexity.

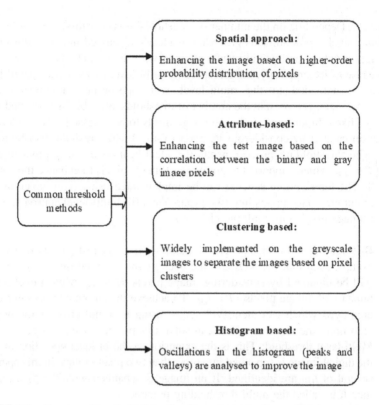

FIGURE 3.2   Common Threshold Selection Procedures for Image Enhancement.

Normally, every thresholding operation needs a monitoring parameter to guide the threshold selection process. Maximization of the monitoring parameter is always considered in the literature to enhance the quality of the test image [14,15]. Though a considerable number of monitoring techniques exist in the literature, few techniques are largely implemented in the image thresholding operation due to its superiority. According to these functions, it can be classified as (i) between-class-based techniques and (ii) entropy-based methods.

Between-class assisted image thresholding was proposed by Otsu and, due to its eminence, a number of research employed Otsu's function to identify the optimal threshold during bi-level and multi-level operations. Further, maximization of entropy-based methods is also adopted in the literature to process a class of images. The subsequent sections present the details on commonly used objective functions in the literature to threshold the chosen image.

3.4 THRESHOLDING FOR GREYSCALE AND RGB IMAGES

Thresholding processes employed for greyscale images work well on RGB images. Only, the threshold selection being implemented for the gray image is based on its

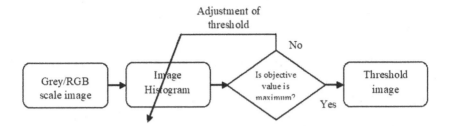

FIGURE 3.3 Optimal Threshold Selection Based on Otsu's Between-Class Variance.

histogram. In the case of the RGB image, threshold selection is implemented separately for its R, G, and B channels.

3.4.1 THRESHOLDING WITH BETWEEN-CLASS VARIANCE

Otsu's technique was discussed in 1979 and it works based on the between-class-variance concept. It identifies the finest threshold by maximizing the objective value [16]. The between-class-variance is Otsu's non-parametric threshold selection concept which is computed by randomly varying the image threshold using a chosen technique. During this process, the selection of $T = t_{OP}$ is essential to convert the raw image into a processed image. This procedure works well on both bi-level and multi-level threshold processes. Its implementation is depicted in Figure 3.3.

Let us consider the greyscale image. During the bi-level operation, $T = t_o$, t_1 are selected, which divides the input image into two groups, such as G_0 and G_1 (image background and ROI). The group G_0 contains the gray pixels of range 0 to t_o and class G_1 encloses the gray levels from t_1 to 255 [17].

This function can be mathematically expressed based on its probability sharing function. Distribution for the gray values G_0 and G_1 can be indicated as follows:

$$G_0 = \frac{P_0}{\eta_0(T)} \cdots \frac{P_{t_0-1}}{\eta_0(T)} \text{ and } G_1 = \frac{P_{t_0}}{\eta_1(T)} \cdots \frac{P_{255}}{\eta_1(T)} \tag{3.1}$$

where $\eta_0(T) = \sum_{i=0}^{T-1} P_i$, $\eta_1(T) = \sum_{i=T}^{255} P_i$

The mean values; Ψ_0 and Ψ_1 for G_0 and G_1 can be denoted as follows:

$$\Psi_0 = \sum_{i=0}^{T-1} \frac{iP_i}{\eta_0(T)} \text{ and } \Psi_1 = \sum_{i=T}^{255} \frac{iP_i}{\eta_1(T)} \tag{3.2}$$

The mean intensity (Ψ_T) of the complete picture can be symbolized as follows:

$\Psi_T = \eta_0 \Psi_0 + \eta_1 \Psi_1$ and $\eta_0 + \eta_1 = 1$

The objective function for the bi-level thresholding problem can be expressed as follows:

$$Otsu_{max} = J(T) = \vartheta_0 + \vartheta_1 \tag{3.3}$$

where $\vartheta_0 = \eta_0 (\Psi_0 - \Psi_T)^2$ & $\vartheta_1 = \eta_1 (\Psi_1 - \Psi_T)^2$

This technique can be modified for a multi-level threshold problem by including various 'T' values as follows:

Choose a test image to be considered which has thresholds distributed as $T = (t_0, t_1, ..., t_{L-1})$ which helps to divide the chosen image into multiple thresholds such as G_0 with gray thresholds 0 to t_0, G_1 with gray thresholds t_0 to t_1, ..., G_T with gray thresholds t_T to 255.

The objective function for multi-level thresholding can be defined as follows:

$$Otsu_{max} = J(T) = \vartheta_0 + \vartheta_1 + ...+ \vartheta_{L-1} \tag{3.4}$$

where $\vartheta_0 = \eta_0 (\Psi_0 - \Psi_T)^2$, $\vartheta_1 = \eta_1 (\Psi_1 - \Psi_T)^2$, ..., $\vartheta_T = \eta_T (\Psi_T - \Psi_{L-1})^2$

Based on the requirement, the threshold value can be chosen as $T = 2, 3, 4, ..., L - 1$.

Advantages: Otsu's technique is a common and frequently used technique to process a variety of images. This approach works well on greyscale images and helps attain better values of image quality parameters compared to other related techniques.

Limitations: Even though this technique works on a variety of images, it performs less compared to the entropy-based technique when the abnormality in the image, which is what needs to be examined, is the prime ROI.

3.4.2 THRESHOLDING WITH ENTROPY FUNCTIONS

Entropy-based assessment of medical images and signals are accepted work in research. This information plays a vital role during abnormality detection.

- **Tsalli's approach**
  Normally, entropy is concerned with the calculation of disorder in an arrangement. Shannon originally utilized the entropy-based assessment to calculate the uncertainty in the sequence of the scheme [18]. Shannon assured that, when a substantial structure is detached as two statistically free subsystems F_1 and F_2, then its entropy can be expressed as [19]:

$$S_h(F_1 + F_2) = S_h(F_1) + S_h(F_2) \tag{3.5}$$

A non-extensive entropy-based concept was introduced by Tsallis from the above equation as shown below:

$$S_{hQ} = \frac{1 - \sum_{i=1}^{T} (P_i)^Q}{Q - 1} \tag{3.6}$$

where T = threshold and Q = entropic index. Equation (3.5) will meet Shannon's entropy when $Q \rightarrow 1$.

The entropy value can be expressed with a pseudo additive rule as:

$$S_{hQ}(F_1 + F_2) = S_{hQ}(F_1) + S_{hQ}(F_2) + (1 - Q) \cdot S_{hQ}(F_1) \cdot S_{hQ}(F_2) \qquad (3.7)$$

Tsalli's technique is then considered to find T of the chosen image. Let the test image have L gray values of span $\{0, 1, 2, ..., L\}$ with probability functions; $P_i = P_0, P_1, ..., P_{L-1}$; for $L = 256$.

In case of the multi-level case, it can be presented as

$$Tsallis\,(t_i) = [t_0, t_1, ...,t_{L-1}] = argmax\ [S_{hQ}^{F_1}(t) + S_Q^{F_2}(t) + ... + S_Q^{k}(t) + (1 - Q) \cdot S_Q^{F_1}(t)$$
$$\cdot S_Q^{F_2}(t), ..., S_Q^{F_K}(t)] \qquad (3.8)$$

where

$$S_{hQ}^{F_1}(t) = \frac{1 - \sum_{i=0}^{t_1-1}\left(\frac{P_i}{P^{F_1}}\right)^Q}{Q-1}, \quad P^{F_1} = \sum_{i=0}^{t_1-1} P_i$$

$$S_{hQ}^{F_2}(t) = \frac{1 - \sum_{i=t_1}^{t_2-1}\left(\frac{P_i}{P^{F_2}}\right)^Q}{Q-1}, \quad P^{F_2} = \sum_{i=t_1}^{t_2-1} P_i \qquad (3.9)$$

$$S_Q^{F_k}(t) = \frac{1 - \sum_{i=t_k}^{L-1}\left(\frac{P_i}{P^k}\right)^Q}{Q-1}, \quad P^{F_K} = \sum_{i=t_k}^{L_2-1} P_i$$

Subject to the following constraints:

$$|P^{F_1} + P^{F_2}| - 1 < S_h < 1 - |P^{F_1} - P^{F_2}|$$
$$|P^{F_2} + P^{F_3}| - 1 < S_h < 1 - |P^{F_2} - P^{F_3}| \qquad (3.10)$$
$$|P^{F_k} + P^{F_{L-1}}| - 1 < Sh < 1 - |P^{F_k} - P^{F_{L-1}}|$$

In multi-level thresholding, the main aim is to discover the best threshold value which maximizes $Tsallis\,(t_i)$.

Advantages: Entropy-assisted technique works well on a class of image cases. Particularly, it offers better results during the image abnormality examination task.

Limitations: Compared to other approaches, the image quality measures attained with Tsallis function is lower, and the grouping of the pixel achieved with this technique is poor compared to Shannon's technique.

- **Fuzzy-Tsallis entropy approach**

Fuzzy-Tsallis (FT) entropy is derived from the Tsallis function, considered to solve the bi-level and multi-level thresholding problem [20,21]. Former work confirms that this approach offers enhanced image quality compared with Tsallis technique.

The mathematical description of FT is expressed as follows:

Let us consider an image (I) of dimension AxB with the maximum threshold value of $L = 256$.

Then $I = \{(i, j): i = 0, 1, \ldots, A - 1; j = 0, 1, \ldots, B - 1\}$ for   thresholds $t = 0, 1, \ldots, L - 1$.

where, A, B & L are positive numbers.

Let, $I_G(X, Y)$ be the gray-level value of the test image at pixel (X, Y) then

$$I_t = \{(X, Y): I_G(X, Y) = t, \ (X, Y) \in I\} \tag{3.11}$$

Let us consider bi-level thresholding for the discussion which separates the image into two sections, such as $I = I_1, I_2$.

For the image (I) the probability distribution can be expressed as follows:

$$\prod_2 = \{I_1, I_2\} \tag{3.12}$$

$$p_3 = P(I_3), \ p_1 = P(I_1) \tag{3.13}$$

where I_1 = darker pixel group (denoted as D or 1), I_2 = brighter pixel group (denoted as B or 2) and the p depicts the probability distribution.

The distribution can be stated for the every t as

$$\left.\begin{array}{l} I_{t1} = \{(X, Y): I_G(X, Y) \le t, (X, Y) \in I_t\} \\ I_{t2} = \{(X, Y): I_G(X, Y) > t, (X, Y) \in It\} \end{array}\right\} \tag{3.14}$$

$$\left.\begin{array}{l} p_{t1} = P(I_{t1}) = p_t * p_{1/t} \\ p_{t2} = P(I_{t2}) = p_t * p_{2/t} \end{array}\right\} \tag{3.15}$$

In Equation (3.15), $p_{1/t}$ and $p_{2/t}$ are conditional probabilities of a pixel which is grouped as dark and bright, with respect to I_t with $p_{1/t} + p_{2/t} = 1 (t = 0, 1, \ldots, 255)$. Let, $\eta_1(t)$ and $\eta_2(t)$ are the grades for dark and bright pixel groups, then

$$\left.\begin{array}{l} p_1 = \Sigma_{t=0}^{255} p_t * p_{1/t} = \Sigma_{t=0}^{255} p_t * \eta_1(t) \\ p_2 = \Sigma_{t=0}^{255} p_t * p_{2/t} = \Sigma_{t=0}^{255} p_t * \eta_2(t) \end{array}\right\} \tag{3.16}$$

The grade $\eta_1(t)$ and $\eta_2(t)$ can be attained using fuzzy membership functions $M_{F1}(t, a, c)$ and $M_{F2}(t, a, b, c)$ correspondingly (Figure 3.4). It is presented in Figure 2.3 and its expression is presented below

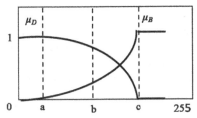

FIGURE 3.4 Fuzzy Membership Function for a Chosen Image.

$$\eta_1(t) = \begin{cases} 1, & t \leq a \\ 1 - \dfrac{(t-a)^2}{(c-a)*(b-a)}, & a < t \leq b \\ \dfrac{(t-c)^2}{(c-a)*(c-b)}, & b < t \leq c \\ 0, & t > c \end{cases} \qquad (3.17)$$

$$\eta_2(t) = \begin{cases} 0, & t \leq a \\ \dfrac{(t-a)^2}{(c-a)*(b-a)}, & a < t \leq b \\ 1 - \dfrac{(t-c)^2}{(c-a)*(c-b)}, & b < t \leq c \\ 1, & t > c \end{cases} \qquad (3.18)$$

where the position of a, b, and c will be between the thresholds; $0 \leq a \leq b \leq c \leq 255$ For every pixel cluster, the FT can be expressed as

$$\left. \begin{aligned} S_1 &= \frac{1 - \sum_{t=0}^{255} (\frac{P_k * \eta_1(k)}{P_1})^Q}{Q - 1} \\ S_2 &= \frac{1 - \sum_{t=0}^{255} (\frac{P_k * \eta_2(t)}{P_2})^Q}{Q - 1} \end{aligned} \right\} \qquad (3.19)$$

Finally, total FT can be represented as conventional Tsallis function:

$$S_Q(1+2) = S_Q(1) + S_Q(2) + (1 - Q) * S_Q(1) * S_Q(2) \qquad (3.20)$$

$$Tsallis_{max}(T) = arg\ max\ [S_Q^1(T) + S_Q^2(T)(1 - Q) \cdot S_Q^1(T) \cdot S_Q^2(T)] \qquad (3.21)$$

Advantages: This procedure works well during the medical image evaluation task and exactly separates the abnormality section from the image background.
Limitations: Implementation of this technique is complex compared to the traditional Tsallis technique.

- **Shannon's approach**
  Description of Shannon's Entropy (SE) was discussed by Kannappan [22]. In SE, let us chose a test-picture of size X * Y. The pixel association in test picture (a, b) is defined as F(a, b), for $x \in \{1, 2, ..., X\}$ and $y \in \{1, 2, ..., Y\}$. Let L be the number of gray levels of the test picture and the set of every gray value $\{0, 1, 2, ..., L - 1\}$ can be denoted as O, in such a way that:

$$F(X, Y) \in O \quad \forall \ (a, b) \in Image \tag{3.22}$$

Then, the normalized histogram will be

$$S = (t_0, t_1, ..., t_{L-1}) \tag{3.23}$$

For a bi-level thresholding case, Equation (3.5) becomes

$$S(T) = a_0(T_1) + a_1(T_2) \tag{3.24}$$

$$E(T) = \max_T \ \{S(T)\} \tag{3.25}$$

where $T = \{T_1, T_2, ..., T_L\}$ is the threshold value, $S = \{a_0, a_1, ..., a_{L-1}\}$ is the normalized histogram, and $_{E(T)}$ is the optimal threshold. This technique for an RGB-scale image is separately implemented for R, G, and B threshold cases. More information on SE can be found in [23].

Advantages: This approach provides better image enhancement results compared to other procedures, presenting better values of image quality measures than Otsu.
Limitations: Implementation and identification of the maximum threshold is quite complex compared to Otsu.

- **Kapur's approach**

Kapur's function has been primarily proposed for thresholding greyscale pictures using the histogram's entropy [24]. This technique finds the optimal threshold by maximizing the entropy.

The threshold vector $T = (t_0, t_1, ... , t_{L-1})$ for a considered image utilizing the Kapur's entropy is given in the following paragraphs:

Choose a dimension of the greyscale image with L gray-levels (0 to L − 1) with a total pixel value of Z, where $f(k)$ represents the frequency of kth intensity-level; then the pixel distribution of the image will be

$$Z = f(0) + f(1) + ... + f(L - 1) \tag{3.26}$$

If the probability of kth intensity-level is given by

$$p(k) = f(k)/Z \tag{3.27}$$

Then during the threshold selection, the pixels of image are separated into $T + 1$ groups according to the assigned threshold value. After extrication of the images as per the selected threshold, the entropy of each cluster is independently calculated and combined to get the final entropy as follows:

$$Bi - level\,threshold = f(t_1, t_2) = e_0 + e_1 \tag{3.28}$$

$$Multilevel\ threshold = f(t_1, t_2, \dots t_L) = e_0 + e_1 + \dots + e_{L-1} \tag{3.29}$$

$$
\begin{aligned}
e_0 &= -\sum_{k=0}^{k=t_1-1} \frac{p_k}{\sigma_0} \ln \frac{p_k}{\sigma_0}, \ \sigma_0 = \sum_{k=0}^{k=t_1-1} p_k \\
e_1 &= -\sum_{k=t_1-1}^{k=t_1-2} \frac{p_k}{\sigma_1} \ln \frac{p_k}{\sigma_1}, \ \sigma_1 = \sum_{k=t_1-1}^{k=t_1-2} p_k \\
e_{L-1} &= -\sum_{k=t_L-1}^{k=t_L-2} \frac{p_k}{\sigma_{L-1}} \ln \frac{p_k}{\sigma_{L-1}}, \ \sigma_{L-1} = \sum_{k=t_L-1}^{k=t_L-2} p_k
\end{aligned}
\tag{3.30}
$$

where e = entropy, p = probability distribution, and σ = probability occurrence.

$$Kapur_{max}(T) = \sum_{p=1}^{L-1} H_j^C \tag{3.31}$$

Other information on Kapur's function can be found in [25].

Advantages: It is a widely considered entropy function and provides better results on a class of greyscale and RGB-scale images.

Limitations: The image quality measures attained with this technique are poor compared to SE.

3.5 CHOICE OF THRESHOLD SCHEME

From the above discussion, it is clear that threshold processes improve the texture and visibility of the image as per the user's need. The selection of the right procedure (Otsu or entropy) and the choice of the threshold value (bi-level or multilevel) follow these considerations:

- The type of the image to be processed (color, dimension, histogram complexity, etc): The literature confirms that Otsu's technique is a common procedure used to process a variety of images with less computation complexity. Next to Otsu, Kapur's function is also implemented. Kapur's technique is needed when the assessment of the image abnormality is a prime task. Tsallis's and Shannon's techniques are chosen to enhance RGB-scale images compared to greyscale images. TE can be implemented where a bi-level thresholding is needed. Further, entropy methods can be used for medical image examination tasks.

- Complexity in image: Simple images with even pixel distribution can be done with Otsu, and for complex cases, an entropy technique can be implemented.

The choice of threshold function depends mainly on prior knowledge. If essential, a detailed comparison among the Otsu and a chosen entropy technique can be performed to confirm the method to be used further. The comparison of the Otsu and Kapur can be found in earlier literature.

3.6 PERFORMANCE ISSUES

Implementation of bi-level thresholding is simple compared to multi-level thresholding. Further, thresholding of a greyscale image is very straightforward compared to RGB. Hence, before implementing a chosen threshold operation, it is essential to verify the following constraints: dimension of the image, pixel distribution, color, and the objective function to be satisfied [26–29].

- Color: The color of the image pixel is the first consideration since greyscale images have a single histogram while RGB-scale images have three, one for each prime pixel (i.e. R, G, B). The thresholding procedure developed for the RGB-scale will work on the greyscale picture, but the approach developed for the greyscale image will not offer the same result with an RGB-scale image. Hence, it is necessary to choose appropriate procedures for RGB images.
- Dimension: The complexity in image enhancement also increases as the dimensions increase. If the dimension is higher, there will be a bigger number of pixels to be examined, which increases the thresholding burden. Hence, in most medical image examination tasks, the image dimension is fixed as $256 \times 256 \times 1$ or $256 \times 256 \times 3$.
- Pixel distribution: Uneven pixel distribution and the image associated with the noise also increases complexity during the thresholding process. Hence, a pre-processing technique is used along with the thresholding process to reduce complexity.

3.7 EVALUATION AND CONFIRMATION OF THRESHOLDING TECHNIQUE

The superiority of the threshold outcome can be assessed based on a comparative analysis between raw image (R) and threshold image (T). During this operation, each pixel of these images are separately compared and, based on its value, the essential image quality parameters are computed. The quality of the T is assessed by well-known image metrics, such as the Root Mean Squared Error (RMSE), Peak Signal-to-Noise Ratio (PSNR), Mean Structural Similarity Index (MSSIM), Normalized Absolute Error (NAE), Normalized Cross-Correlation (NCC), Average Difference (AD), and Structural Content (SC) [30,31].

Mathematical expression of the considered image quality measures are presented below:

$$PSNR_{(R,T)} = 20 \, log_{10} \left(\frac{255}{\sqrt{MSE_{(R,T)}}} \right) dB \qquad (3.32)$$

$$RMSE_{(R,T)} = \sqrt{MSE_{(R,T)}} = \sqrt{\frac{1}{XY} \sum_{i=1}^{X} \sum_{j=1}^{Y} [R_{(i,j)} - T_{(i,j)}]^2} \qquad (3.33)$$

The mean SSIM is generally used to estimate the image superiority and inter dependencies between the original and processed image.

$$MSSIM_{(R,T)} = \frac{1}{M} \sum_{z=1}^{M} SSIM_{(R_z,T_z)} \qquad (3.34)$$

where R_z and T_z are the picture contents at the zth local window, and M is the number of local windows in the picture.

$$NAE_{(R,T)} = \frac{\sum_{i=1}^{X} \sum_{j=1}^{Y} \left| R_{(i,j)} - T_{(i,j)} \right|}{\sum_{i=1}^{X} \sum_{j=1}^{Y} \left| R_{(i,j)} \right|} \qquad (3.35)$$

$$NCC_{(R,T)} = \frac{\sum_{i=1}^{X} \sum_{j=1}^{Y} R_{(i,j)} \cdot T_{(i,j)}}{\sum_{i=1}^{X} \sum_{j=1}^{Y} R_{(i,j)}^2} \qquad (3.36)$$

$$AD_{(R,T)} = \frac{\sum_{i=1}^{X} \sum_{j=1}^{Y} R_{(i,j)} - T_{(i,j)}}{XY} \qquad (3.37)$$

$$SC_{(R,T)} = \frac{\sum_{i=1}^{X} \sum_{j=1}^{Y} R_{(i,j)}^2}{\sum_{i=1}^{X} \sum_{j=1}^{Y} T_{(i,j)}^2} \qquad (3.38)$$

In all the expressions, $X \times Y$ represents the size of the considered image, R is the original test image, and S is the segmented image of a chosen threshold. A higher value of PSNR, MSSIM, NCC and lower value of RMSE, NAE, AD, SC specifies a superior quality of thresholding. Improved fitness function with minor CPU time during the optimization search also confirms the capability of the considered heuristic algorithm.

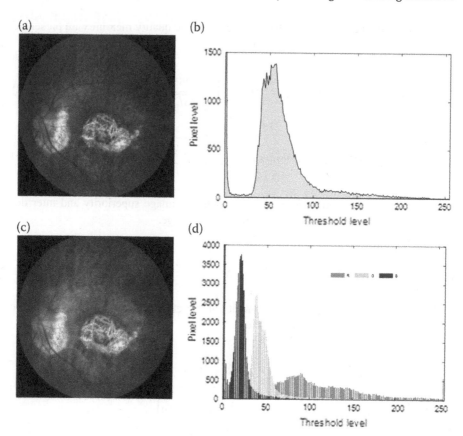

FIGURE 3.5 Retinal (Gray/RGB) Test Image and the Related Histograms.

3.8 THRESHOLDING METHODS

In the literature, a number of picture thresholding techniques are projected and executed on the RGB/greyscale images of a range of dimensions. Identification of the threshold value for greyscale illustration is fairly easy compared to the RGB-scale images since, in the RGB scale picture, the pixel distribution is fairly complex and each pixel is the mixtures of R, G, and B pixel points. The recognition of the optimum threshold is to be realized independently for the R-, G- and B-channels. Complication rises based on the pixel level and its non-linear allocation. In the case of the greyscale image, the pixel distribution is even and identification of the optimum threshold is simple [32–35].

Let us consider a fundus test image for the discussion and the dimension of the test image is chosen as 256 × 256 × 3 pixels (for RGB) and 256 × 256 × 1 pixels (for gray scale) for the demonstration.

Figure 3.5 presents the fundus retinal image considered for the experimental demonstration and in which Figure 3.5 (a) and (c) presents the test images and Figure 3.5 (b) and (d) presents the related histograms. It can be noted from this illustration that the histogram of RGB-scale images is relatively non-linear and difficult compared to greyscale threshold.

In this work, histogram-assisted thresholding is considered and the selection of the histogram is achieved through a selected methodology existing in the literature. During this exploration, the employed system is allowed to adjust the thresholds of the picture until the assigned objective function is reached. The thresholding procedure can be implemented with the traditional (operator-assisted method) practice or the computerized (heuristic algorithm assisted) method. The selection mainly depends on the knowledge of the operator and the complexity of the thresholding practice.

3.9 RESTRICTIONS IN TRADITIONAL THRESHOLD SELECTION PROCESS

Image threshold is one of the commonly used illustration enhancement schemes in which a preferred approach is used to improve the examined image based on the pixel grouping concept. The identification of finest threshold is fairly simple when the implemented division is bi-level (Test image = SOI + Background), which can be implemented with a manual operator when the image is in greyscale. The computation time needed for a manual operator is moderately longer. Due to the complexity of RGB-scale images and the multi-level threshold selection, heuristic algorithm-based techniques are widely implemented. Next, the histogram of the retinal image is considered as shown in Figure 3.6 (a) and (b), where Figure 3.6 (b) needs to be detached into three pixel groups.

It is clear from Figure 3.6 (a) and (b) that the gray pixel distribution is homogeneous from thresholds 0 to 255. Finding the necessary threshold assessment is relatively complex. Further, the pixel closer to zero threshold is darker (background) and the pixel closer to the threshold 255 is very light. The problem chosen in this study is to identify the threshold for various pixel groups, (i.e. Figure 3.6) for both the gray- and RGB-scale histograms.

If grouping of pixels is implemented manually, a multitude of threshold combinations need to be tried and the attained outcome is to be verified to substantiate the superiority of the threshold process. This procedure is extremely tedious for RGB-scale imaging thus, in recent years, heuristic algorithm based-techniques have been widely employed to identify the optimal threshold for the test image by maximizing the between-class or a chosen entropy function.

Heuristic algorithm-assisted procedures are currently used in a range of applications to pre-process and post-process images with the most successful methods existing in the literature. These algorithms reduced the complexity in the image thresholding problem considerably and work well on various images, even though it comes with irregularities.

3.10 NEED FOR HEURISTIC ALGORITHM

Heuristic and meta-heuristic algorithms (HA) are proposed using the mathematical models of some well-known, trouble-solving capabilities found in the literature. Due to the progression in computing technology, HA can easily be executed to resolve a selection of constrained and unconstrained afflictions existing in the

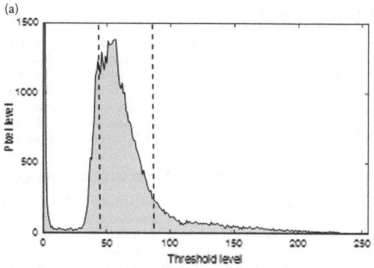

Multi-thresholding for grayscale histogram

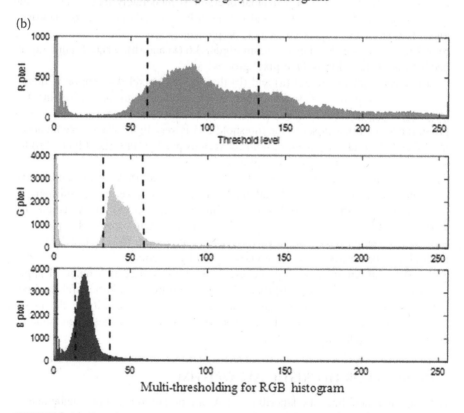

Multi-thresholding for RGB histogram

FIGURE 3.6   Grouping of the Image Pixels during the Multi-Threshold Process.

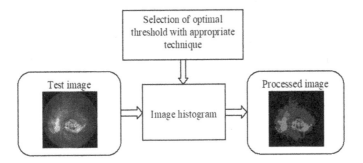

FIGURE 3.7   Traditional Threshold Selection Process.

world. This chapter presents the outline of the heuristic algorithms commonly used in the image processing literature.

Figure 3.7 presents the frequently used histogram-based threshold problem where the histogram of the raw image is explored by the chosen HA to get the processed image. To demonstrate this, benchmark image with a pseudo name Bird is considered. This image shows the output of a bi-level threshold operation in which the ROI is exactly separated with respect to the background.

Advantages: The HA practice works fine for a class of images irrespective of their difficulty. Further, this practice helps decrease the computational burden of the multi-thresholding process.

Limitations: Identification of an appropriate HA is a moderately challenging task and this practice needs preliminary tuning for the algorithm parameters.

3.11 SELECTION OF HEURISTIC ALGORITHM

The thorough understanding of the HA-based threshold choice is depicted in Figure 3.8. The different stages concerned in this maneuver includes (i) selecting an appropriate HA and regulating its constraint based on the problem to be solved; (ii) considering the Objective-Function (OF) to be maximized throughout the threshold choice; (iii) randomly altering the histogram of the image until the maximized OF is attained; (iv) comparing the threshold image with the trial image and computing the necessary Picture-Quality-Measures (PQM); and (v) validating the applied practice based on the attained PQM.

From the discussion, it is clear that the option for a particular HA plays a foremost role in thresholding. In this work, the established heuristic procedures such as Particle-Swarm Optimization (PSO), Bacterial-Forging Optimization (BFO), Firefly-Algorithm (FA), Bat-Algorithm (BA), Cuckoo-Search (CS), Social-Group optimization (SG), Teaching-Learning-Based Optimization (TLBO) and Jaya Algorithm (JA) are considered for the appraisal.

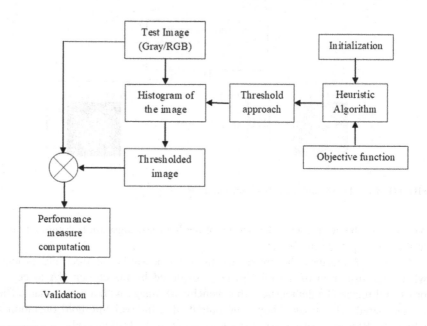

FIGURE 3.8   Multi-Level Threshold Selection Using Heuristic Algorithm.

3.11.1 PARTICLE SWARM OPTIMIZATION

PSO, proposed by Kennedy and Eberhart in 1995 [36], is a universal optimization exercise created with inspiration from a community performance found in bird and fish groups. Because of its performance, it is expansively practical in a selection of domains thanks to its elevated computational efficiency. In contrast to previous population-based stochastic systems, PSO has equal or even improved exploration performance for numerous hard optimization exertions, with faster and steadier convergence rates. It has been established to be a winning optimization procedure in different illustration-processing domains [37,38].

In PSO, the amount of modification constraints is less compared to other techniques. Here, a group of virtual birds are initialized with random locations L_i and speeds S_i. At the early searching stage, each bird in the cluster is spread randomly throughout the d sized examination space. With the management of the OF and flying information, each bird in the group energetically adjust its flying position and speed. During the investigation, every particle memorizes its best position reached so far (i.e. $p_{\text{best}} - (P_{i,d}^t)$) while also collecting the universal best location information accomplished by any particle in the population (i.e. $g_{\text{best}} - (G_{i,d}^t)$).

At iteration t, every particle i has its position defined by $L_{i,n}^t = [L_{i,1}, L_{i,2}, ..., L_{i,d}]$ and speed distinct as $S_{i,n}^t = [S_{i,1}, S_{i,2}, ..., S_{i,d}]$ in exploring space d. Speed and position of each particle in the following iteration can be calculated as:

$$S_{i,d}^{t+1} = W * S_{i,d}^t + q_1 * r_1 * (P_{i,d}^t - L_{i,d}^t) + q_2 * r_2 * (G_{i,d}^t - L_{i,d}^t) \quad (3.39)$$

$$S_{i,d}^{t+1} = \Psi * [S_{i,d}^t + q_1 * r_1 * (P_{i,d}^t - L_{i,d}^t) + q_2 * r_2 * (G_{i,d}^t - L_{i,d}^t)] \quad (3.40)$$

where $i = 1, 2, ..., k$ and $N = 1, 2, ..., d$

$$L_{i,N}^{t+1} = \begin{cases} L_{i,N}^t + S_{i,N}^{t+1} \ if \ L_{mini,N} \le L_i^{t+1} \le L_{maxi,N} \\ L_{mini,N} \ if \ L_{i,N}^{t+1} < L_{mini,N} \\ L_{mani,N} \ if \ L_{i,N}^{t+1} > L_{maxi,N} \end{cases} \quad (3.41)$$

In Equation (3.42), the inertia of weight W symbolizes a significant factor for PSO's convergence. It is used to manage the impact of preceding velocities on the present velocity. An inertia weight factor facilitates global exploration whereas minute weight factor facilitates local exploration. Thus, it is better to choose large weight factors for preliminary iterations and to gradually diminish the weight factor in consecutive iterations. This can be done with the following equation:

$$w = w_{max} - (w_{max} - w_{min}) * Iter/Iter_{max} \quad (3.42)$$

where w_{max} and w_{min} are initial and ending weight, respectively, $Iter$ is iteration amount and $Iter_{max}$ is the maximum iteration.

In Equation (3.43), constriction value Ψ is dependable for modernizing the speed of PSO which fixes convergence and accuracy. This factor can be assigned using the following equation:

$$\Psi = \frac{2}{\left| 2 - \beta - \sqrt{\beta^2 - 4\beta} \right|};$$

$$\text{where } \beta = q_1 + q_2, \beta > 4 \quad (3.43)$$

Acceleration invariable q_1, called the cognitive limit, pulls each element towards the best local spot while constant q_2, called the social limit, pulls the particle near the best global spot. r_1 and r_2 are known as arbitrary numbers in the range 0–1. The particle position is modified by Equation (3.40). This process is repeated until the stopping criterion is reached. The flow chart of the PSO is presented in Figure 3.9 [39].

3.11.2 BACTERIAL FORAGING OPTIMIZATION

BFO is a type of biologically-motivated stochastic exploring practice based on mimicking the foraging (methods for locating, handling and ingesting food) actions of *Escherichia coli* (*E. coli*). Due to its qualities such as efficiency, easy implementation, and stable convergence, it is widely applied to solve a range of complex engineering optimization problems [40–43].

In BFO, an OF is created as the endeavour for the bacteria in search of food. In this algorithm, a set of simulated bacteria tries to achieve optimum strength through

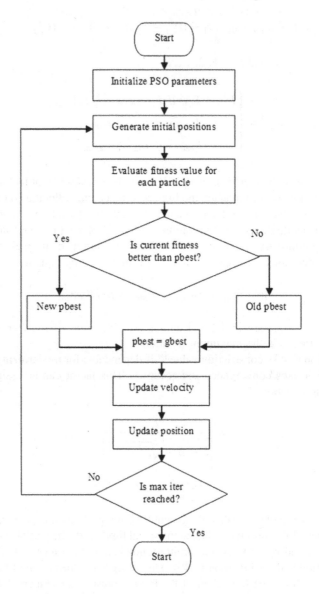

FIGURE 3.9   Flow Chart of the PSO-Based Optimization.

the phases of chemotaxis, swarming, reproduction, elimination, and dispersal. Each bacterium produces a result iteratively for a set of optimal values. Slowly, all the microbes meet the global optimum. In the chemotaxis stage, the bacteria either resorts to a tumble, run, or swim. During swarming, each *E. coli* bacterium signals another bacterium such that the attractants swarm mutually. In addition, during reproduction, the least healthy bacteria die among the healthiest, each bacterium splits into two bacteria, which are placed at the same location. While in the

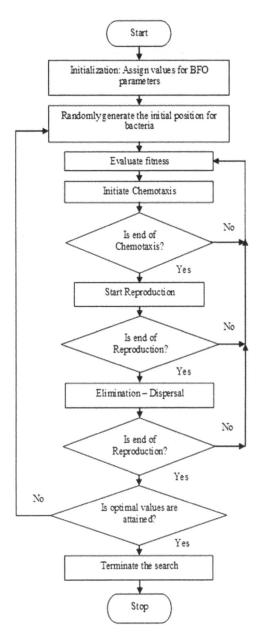

FIGURE 3.10   Flow Chart of Traditional BFO Algorithm.

elimination and dispersal stages, any bacterium from the total set can either be eliminated or dispersed to a random location during the optimization. This stage prevents the bacteria from attaining the local optimum. The flow chart of the BFO algorithm is depicted in Figure 3.10.

3.11.3 FIREFLY ALGORITHM

FA is one of the most successful HA. The principal parameters which choose the effectiveness of the FA are the difference of light strength and attractiveness linking adjacent fireflies. These two parameters will be affected by an increase in the distance between fireflies. An introduction can be found in [44–46].

Change in luminance can be systematically articulated with the following Gaussian form:

$$I(r) = I_0 e^{-\gamma d^2} \tag{3.44}$$

where I = fresh brightness, I_0 = original brightness, and γ = light fascination coefficient.

The attractiveness towards the luminance can be analytically represented as:

$$\gamma = \gamma_0 e^{-\beta d^2} \tag{3.45}$$

where γ = attractiveness coefficient, and γ_0 = attractiveness at $r = 0$.

The above expression explain a quality space $\Gamma = 1/\sqrt{\beta}$ over which the attractiveness changes considerably from γ_0 to $\gamma_0 e - 1$. The attractiveness function $\gamma(d)$ can be any monotonically falling functions such as

$$\gamma(d) = \gamma_0 e^{-\beta d^q}, \ (q \geq 1) \tag{3.46}$$

For a fixed γ, the characteristic length becomes

$$\Gamma = \beta^{-1/q} \rightarrow 1, q \rightarrow \infty \tag{3.47}$$

Conversely, for a given length scale Γ, the parameter γ can be used as atypical initial value (that is $\beta = 1/\Gamma q$).

The Cartesian space among two fireflies a and b at d_a and d_b, in the G dimensional exploring space can be scientifically expressed as

$$G_{ab}^t = \|d_b^t - d_a^t\|_2 = \sqrt{\sum_{k=1}^{n} (d_{b,k} - d_{a,k})^2} \tag{3.48}$$

In FA, the luminosity at a particular space G from the light source d_a^t obeys the inverse square law. The light intensity of a firefly I, as the distance u increases in terms of $I \propto 1/u^2$. The movement of the attracted firefly a towards a brighter firefly b can be determined by the following position updated equation:

$$d_a^{t+1} = d_a^t + \beta_0 e^{-\gamma u_{ab}^2}(d_b^t - d_a^t) + \Re \tag{3.49}$$

where, d_a^{t+1} = updated position of the firefly, d_a^t = initial position of the firefly, $\beta_0 e^{-\gamma u_{ab}^2}(d_b^t - d_a^t)$ = attraction between fireflies, and $\Re$ = randomization parameter.

From Equation (3.49), it is observed that the updated position of the ith firefly depends on its initial position, attraction of fireflies toward the luminance, and the randomization parameter. Other information on FA can be found in [47–49].

3.11.4 BAT ALGORITHM

The BA was proposed in 2010 by Yang [50] to discover the finest answer for arithmetic problems and, due to its worth, recently has been considered to solve various optimization problems. The process of conventional BA was connected with echolocation/bio-sonar characteristic of microbats and is arithmetically represented to construct the BA.

The BA has the following mathematical expressions:

$$\text{Velocity modernization} = S_i^{k+1} = S_i^k + [x_i^k - G_{best}]f_i \tag{3.50}$$

$$\text{Location update} = x_i^{k+1} = x_i^k + S_k^{k+1} \tag{3.51}$$

$$\text{Frequency alteration} = f_i = f_{min} + (f_{max} - f_{min})\alpha \tag{3.52}$$

where α is an arbitrary value of choice [0,1].

Equation (3.52) controls Equations (3.50) and (3.51), hence, the choice of the frequency value should be appropriate.

Updated value for every bat is produced based on

$$x_{new} = x_{old} + \varepsilon A^k \tag{3.53}$$

where ε is an arbitrary value of choice [-1,1] and A = loudness constraint during the exploration.

The expression of the loudness variation can be represented as

$$A_i^{n+1} = \phi A_i(k) \tag{3.54}$$

where ϕ is a variable with a value $0 < \phi < 1$.

Other values on the BA can be found in the literature [51,52].

The conventional working of the BA is shown in Figure 3.11.

The objective of the BA is to recognize the optimal answer for a specified problem by exploring the search area. If a single bat identifies the optimal solution, then it will invite the other bats towards the solution. The probability of getting the global maxima in BA is better compared to other algorithms, thus it is a successful approach used in a variety of optimization tasks.

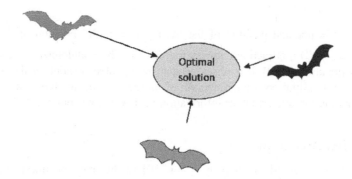

FIGURE 3.11 Optimal Threshold Identification with BA.

3.11.5 CUCKOO SEARCH

The conventional CS was invented by Yang and Deb in 2009 [53] and, after more than a decade, it has become widely used to resolve a variety of image processing functions. In the literature, a substantial amount of development procedures are planned to extend the optimization exploration of the CS.

The conventional CS is developed by mimicking the reproduction trick followed by freeloading cuckoos. The CS is developed by considering the subsequent hypothesis: (i) each cuckoo leaves an egg in an arbitrarily selected nest of the host bird, (ii) the nest with distinguished enduring egg will be carried to the succeeding invention, and (iii) for a selected threshold task, the quantity of host-bird's nest is fixed and it might recognize the cuckoo egg with a probability $P_a \in [0, 1]$. When the host identifies the egg, it may abolish the egg or discard the present nest and the host will construct a new nest.

The above hypothesis is accounted to generate the arithmetical model of the CS. The implementation steps in CS are quite straightforward compared to other existing approaches, and the proposed CS is depicted in Equation (3.55).

$$P_i^{n+1} = P_i^n + \sigma \oplus LF \qquad (3.55)$$

where, P_i^n = early location, P_i^{n+1} = updated location, σ = step size (chosen as 1.2), $\oplus$ = entrywise multiplication, and LF = Levy-Flight operator.

The pseudo-code for the CS is depicted below and its complete details can be found in [54]:

3.11.6 SOCIAL GROUP OPTIMIZATION

SG is an HA developed by Satapathy and Naik [55]. It was formed by replicating the performance and information conveying followed during human grouping. The SG includes two chief functions, namely the (i) improving phase, which synchronizes the location of citizens (agents) based on the OF, and the (ii) acquiring phase that authorizes the agents to discover the optimum potential reply for the problem under discussion.

Initialize the CS with; N = *no. of nests,* $Iter_{max}$ = *maximum iterations,* $P_n \in [0, 1]$, $F(P)$ = *Objective function,* **and** *Stopping constra int* $(F_{max}(P)$ *or* $Iter_{max})$

Initialize the counter (Set $n = 0$**)**

for $(i = 1: I \leq N)$ **do**

                Begin the population of N-host P_i^n

                Appraise $F(P_i^n)$

end for

repeat

                Produce X_i^{n+1} using Equation (1)

                Appraise $F(P_i^{n+1})$

                Select a nest P_j arbitrarily from N-solutions

if $\{F(P_i^{n+1})\}>\{F(P_j^n)\}$ **then**

                Restore P_j with P_i^{n+1}

end if

                Discard the nest based on $P_a \in [0, 1]$

                Construct a new nest

                Keep the optimum solutions (nest with finest solutions)

                Grade and arrange the solutions to discover the finest one

Enhance iteration count (Set $n = n + 1$**)**

repeat till stopping constra int reached $(F_{max}(P)$ *or* $Iter_{max})$

Generate the optimised result.

The arithmetical replica for the SGO is as follows [56–58]:

Let's consider K_i as the preliminary information of people in an assembly and $i = 1$, 2, 3, ..., P, with P as the sum of people in the grouping. If the optimization task needs a D-dimensional exploration space, then the knowledge term can be expressed as $K_i = (k_{i1}, k_{i2}, k_{i3}, ..., k_{id})$. The fitness value for any task can be defined as f_j, with $j = 1, 2, ..., P$. Thus, for the maximization problem, the fitness value can be written as:

$$Gbest_j = max \{f(K_i) \, for \, i = 1, \, 2, \, ..., P\} \qquad (3.56)$$

The steps of the normal SG algorithm can be illustrated as following:

__Standard SG Optimization Algorithm__

__Start__

    __Assume__ five agents $(i = 1,2,3,4,5)$

    __Assign__ these agents to determine the $Gbest_j$ in a D-dimensional investigation space

    __Randomly__ distribute the entire agents in the group all over the investigation space during initialization practice

    __Compute__ the fitness cost based on the task to be solved

Update the direction of agents with $Gbest_j = \max \{f(K_i)\}$

Initiate the improving stage to update the information of supplementary agents in order to achieve the $Gbest_j$

Initiate the acquiring stage to proceed the information of agents by randomly choosing the agents with finest fitness

Repeat the process till the whole agents travel towards the finest potential position in the D-dimensional investigation space

If all the agents contain roughly similar strength values ($Gbest_j$)

Then

conclude the search and exhibit the optimized outcome for the selected task

Else

Repeat the preceding steps

End

Stop

To modernize the location (information) of each individual in the cluster, the improving phase considers the following relation:

$$K_{new_{i,j}} = c * K_{old_{i,j}} + R * (Gbest_j - K_{old_{i,j}}) \tag{3.57}$$

where K_{new} = new information, K_{old} = old information, $Gbest$ = global best information, R = arbitrary number $[0,1]$, and c represents the self-introspection constraint $[0,1]$ with a chosen value of 0.2.

Throughout the acquiring stage, the agents will discover the global result based on information updating practice by arbitrarily choosing one person from the group (K_r) based on $i \neq r$. Once the fitness value becomes $f(K_i) < f(K_r)$, then the following information process is executed:

$$K_{new_{i,j}} = K_{old_{i,j}} + R_a * (K_{i,j} - K_{r,j}) + R_b * (Gbest_j - K_{i,j}) \tag{3.58}$$

where R_a and R_b are random numbers having the range $[0,1]$ and $K_{r,j}$ is the information (position) value of the selected person.

3.11.7 TEACHING-LEARNING-BASED OPTIMIZATION

TLBO is one of the meta-heuristic schemes developed by Rao [59] and is based on the arithmetical replica of the teaching-learning exercise active in the classroom. It is a best possible investigative process amongst two closed linked sets such as teacher and learner. The teacher phase starts learning from a most exceptional teacher and the scholar phase approves the students to be taught during interactions. TLBO is extensively implemented by researchers to resolve a selection of industrial optimization problems. A complete explanation about the TLBO can be found in [59,60].

Pseudo code of the TLBO algorithm

START;
 Initialize the algorithm with amount of learners (N_T), investigate aspect (D), Maximum iteration (M_{iter}) and the objective value (J_{max});
 Randomly initialize 'N_T' learners for x_j (j= 1, 2, ... n);
 Evaluate the practice and decide the optimum solution f(x_{best});
 WHILE iter = 1:M_{iter};

 %TEACHER STAGE%

 Use f(x_{best}) as teacher;
 Sort based on f(x), decide innovative teachers based on:
 f(x)$_s$ = f(x_{best}) − rand for f(x)$_s$ = 2,3,..., T;

 FOR j = 1:n
 Compute $T_F^j = round\,[1 + rand\,(0,\ 1)\{2 - 1\}]$;
 $x_{new}^j = x^j + rand\,(0,\ 1)\,[x_{teacher} - (T_F^j.\ x_{mean})]$;
 % Compute objective value; $f(x_{new}^j)$%
 If f(x_{new}^j) < f(x_j), then $x^j = x_{new}^j$;

 End If % End of TEACHER PHASE%

 %STUDENT PHASE%

 Arbitrarily choose the learner x^j, such that $k \neq j$;
 If $f(x^j) < f(x^k)$, then $x_{new}^j = x^j + rand(0,1)(x^j - x^k)$;
 Else $x_{new}^j = x^j + rand(0,1)(x^k - x^j)$;
 End If
 If x_{new}^j is better than x^j, then $x^j = x_{new}^j$;
 End If % End of STUDENT PHASE%

 End FOR

 Set m = m+1;

 End WHILE

 Record the threshold values, J_{max}, and performance measures and CPU time;
 STOP;

3.11.8 Jaya Algorithm

JA is an evolutionary algorithm by Rao [61]. Its theory and working principle can be found in [62,63]. The chief merit of the JA compared with other considered heuristic/evolutionary approaches is that it involves the smallest premature constraints to be assigned.

Let $G(x)_{max}$ be the preferred objective function, $\Re 1 = \Re 2$ are prejudiced quantities of alternative [0,1], N is population quantity (i.e. $n = 1,2, ..., N$) and i and D designate the quantity of iterations and dimensions (i.e. $j = 1,2, ..., D$) respectively.

At some occurrence in the algorithm, $G(x)_{best}$ and $G(x)_{worst}$ indicate the premium and nastiest results attained during the contestant in population.

The prime equation of JA is presented below

$$X^1_{j,n,i} = X_{j,n,i} + \Re 1_{j,i}(X_{jbest,i} - \left| X_{j,n,i} \right|) - \Re 2_{j,i}(X_{jworst,i} - \left| X_{j,n,i} \right|) \quad (3.59)$$

where, $X_{j,n,i}$ signify the jth variable of nth contestant at ith iteration and $X^1_{j,n,i}$ indicate the modernized value. In this work, JA is utilized to discover the maximized value of the Otsu's between-class variance value for a chosen 'Th'.

Additional details concerning the conventional JA and its recent advancements can be found in recent literature [62,63].

3.12 INTRODUCTION TO IMPLEMENTATION

The execution of a selected thresholding procedure is evidently depicted in Figure 3.8 and this practice will work on a class of conventional and therapeutic images. The principal aim of this method is to utilize the operation on a selected picture to recognize the optimum threshold to grouping image pixels. During execution, the following events are considered:

Step 1: Gathering the illustration to be examined and improving the picture excellence and dimension, if essential.

(This process helps renovate the constriction in the picture and offer a good quality test image.)

Step 2: Choosing the suitable OF and HA to accomplish optimal results.

(This practice helps make a decision on Otsu/Entropy-based method to apply thresholding.)

Step 3: Fixing the amount of threshold needed to improve the SOI of the picture for additional appraisal.

(The threshold value will be assigned as 2 for bi-level process and threshold > 2 for the multi-level process.)

Step 4: Performing the threshold practice and doing again it over a predefined period. Apply the statistical examination to substantiate the strength of the proposed method.

Step 5: After attaining the outcome, computing the PQM.

Step 6: Substantiating the advantages of the executed practice and authenticating the results.

3.13 MONITORING PARAMETER

Normally, HA-based multi-level thresolding is an automatic method performed by means of a devoted computer algorithm. Due to its practicality, a number of illustration processing applications adopted the HA. The excellence and the amount of results depended mostly on the supervising functions employed to manage the automated process. In most cases, a cautiously assigned OF is used as the controller for the whole illustration thresholding process. Due to this, it is necessary to have a better objective function to observe and control the computerized algorithms.

The selection of the OF depends mainly on the practice implemented to execute the process. If the Otsu's function is implemented, the maximized value of the between-class-variance acts as the OF and, in the case of the Tsallis, FT, Shannon, and Kapur, the maximised entropy acts as the OF. Either a single or multiple OF can be utilized. The upcoming section sketches the implantations of the OF for a selected process.

3.13.1 OBJECTIVE FUNCTION

OF plays a leading role in picture thresholding and its selection depends on the implemented practice.

Some frequently used OF along with its processes are represented below

- Otsu's between-class variance: If this function is implemented the OF will be the maximization of the Otsu's constraint depicted as follows:

$$Otsu_{max} = J(T) = \vartheta_0 + \vartheta_1 + ... + \vartheta_{L-1} \tag{3.60}$$

where $\vartheta_0 = \eta_0(\Psi_0 - \Psi_T)^2$, $\vartheta_1 = \eta_1(\Psi_1 - \Psi_T)^2$, ... , $\vartheta_T = \eta_T(\Psi_T - \Psi_{L-1})^2$

- Tsalli's entropy: In this, the entropy of the histogram is considered as the OF and the mathematical expression is depicted as follows:

$$Tsallis\,(t_i) = [t_0, \ t_1, \ \ldots \ ,t_{L-1}] = argmax\,[S_{hQ}^{F_1}(t) + S_Q^{F_2}(t) + \ldots + S_Q^{k}(t)$$

$$+ (1 - Q) \cdot S_Q^{F_1}(t) \cdot S_Q^{F_2}(t), \ \ldots, S_Q^{F_k}(t)] \qquad (3.61)$$

- FT entropy: In this, the entropy of the histogram is considered as the OF and the mathematical expression is depicted as follows:

$$Tsallis_{max}(T) = argmax\,[S_Q^1(T) + S_Q^2(T)(1 - Q) \cdot S_Q^1(T) \cdot S_Q^2(T)] \qquad (3.62)$$

- Shannon's entropy: In this, the entropy of the histogram is considered as the OF and the mathematical expression is depicted as follows:

$$E\,(T) = \max_T \{S\,(T)\} \qquad (3.63)$$

- Kapur's entropy: The entropy of the histogram is considered as the OF and the mathematical expression is depicted as follows:

$$Kapur_{max}(T) = \sum_{p=1}^{L-1} H_j^C \qquad (3.64)$$

3.13.2 Single and Multiple Objective Function

The quality of the threshold image depends on the OF and, based on what is necessary, either single or multiple objective functions can be implemented.

The expressions depicted in Equations (5.1) to (5.5) present the information on single OF. The multiple OF is the modified version of the single OF and, for demonstration, this section considers the Otsu's function. A similar technique can be used for other entropy functions.

$$OF_{single} = Otsu_{max} = J\,(T) = \vartheta_0 + \vartheta_1 + \ldots + \vartheta_{L-1} \qquad (3.65)$$

$$OF_{multiple1} = (W_1 * Otsu_{max}) + (W_2 * PSNR) \qquad (3.66)$$

$$OF_{multiple2} = (W_1 * Otsu_{max}) + (W_2 * PSNR) + (W_3 * SSIM) \qquad (3.67)$$

where W_1, W_2, and W_3 are weighting functions, whose values are assigned as [0,1].

Equation (3.65) depicts the single OF and Equations (3.66) and (3.67) present the multiple OF.

3.14 SUMMARY

This chapter presented the essential thresholding procedures existing in the literature and its practical significance in medical applications. The existing thresholding procedure and the execution of bi-level and multi-level thresholding with between-class-variance and entropy-assisted techniques were also discussed. Further, the need for single and multiple objectives were illustrated with necessary equations. The outline of some well-known heuristic procedures and their applications in multi-thresholding is also presented. This section also laid out the implementation of the thresholding process on a chosen test image and the performance measures considered confirms the superiority of the implemented technique.

REFERENCES

1. Priya, E. & Srinivasan, S. (2016). Automated object and image level classification of TB images using support vector neural network classifier. *Biocybernetics and Biomedical Engineering*, 36(4), 670–678.
2. Priya, E. & Srinivasan, S. (2016). Validation of non-uniform illumination correction techniques in microscopic digital TB images using image sharpness measures. *International Journal of Infectious Diseases*, 45(S1), 406.
3. Priya, E. & Srinivasan, S. (2015). Separation of overlapping bacilli in microscopic digital TB images. *Biocybernetics and Biomedical Engineering*, 35, 87–99.
4. Priya, E., & Srinivasan, S. (2015). Automated identification of tuberculosis objects in digital images using neural network and neuro fuzzy inference systems. *Journal of Medical Imaging and Health Informatics*, 5, 506–512.
5. Priya, E., Srinivasan, S. & Ramakrishnan, S. (2014). Retrospective non-uniform illumination correction techniques in microscopic digital TB images. *Microscopy and Microanalysis*, 20(5), 1382–1391.
6. Bhandary, A., Prabhu, G.A., Rajinikanth, V., Thanaraj, K.P., Satapathy, S.C., Robbins, D.E., Shasky, C., Zhang, Y.D., Tavares, J.M.R.S. & Raja, N.S.M. (2020). *Pattern Recognition Letters*, 129, 271–278.
7. Rajinikanth, V., Dey, N., Raj, A.N.J., Hassanien, A.E., Santosh, K.C. & Raja, N.S.M. (2020). Harmony-search and Otsu based system for coronavirus disease (COVID-19) detection using lung CT scan images, *arXiv preprint*, arXiv:2004.03431.
8. Dey, N., et al. (2019). Social-group-optimization based tumor evaluation tool for clinical brain MRI of Flair/diffusion-weighted modality. *Biocybernetics and Biomedical Engineering*, 39(3), 843–856.
9. Raja, N.S.M., Rajinikanth, V., Fernandes, S.L. & Satapathy, S.C. (2017). Segmentation of breast thermal images using Kapur's entropy and hidden Markov random field. *Journal of Medical Imaging and Health Informatics*, 7(8), 1825–1829.
10. Dey, N., Shi, F. & Rajinikanth, V. (2020). Image examination system to detect gastric polyps from endoscopy images. *Information Technology and Intelligent Transportation Systems*, 323, 107–116.
11. Fernandes, S.L., Rajinikanth, V. & Kadry, S. (2019). A hybrid framework to evaluate breast abnormality using infrared thermal images. *IEEE Consumer Electronics Magazine*, 8(5), 31–36.
12. Nair, M.V., Gnanaprakasam, C.N., Rakshana, R., Keerthana, N. & V Rajinikanth, V. (2018). *International Conference on Recent Trends in Advance Computing (ICRTAC)*, *IEEE*, 174–179.

13. Rajinikanth, V., Raja, N.S.M. & Arunmozhi, S. (2019). ABCD rule implementation for the skin melanoma assesment–A study. In: *IEEE International Conference on System, Computation, Automation and Networking (ICSCAN)*, 1–4.
14. Lakshmi, V.S., Tebby, S.G., Shriranjani, D. & Rajinikanth, V. (2016). Chaotic cuckoo search and Kapur/Tsallis approach in segmentation of *T. cruzi* from blood smear images. *The International Journal of Computer Science and Information Security (IJCSIS)*, 14(CIC 2016), 51–56.
15. Rajinikanth, V., Satapathy, SC., Dey, N., Fernandes, SL. & Manic, KS. (2019). Skin melanoma assessment using Kapur's entropy and level set—A study with bat algorithm. *Smart Innovation, Systems and Technologies*, 104, 193–202.
16. Otsu, N. (1979). A threshold selection method from gray-level histograms. *IEEE Transactions on Systems, Man, and Cybernetics*, 9(1), 62–66.
17. Raja, N.S.M., Rajinikanth, V. & Latha, K. (2014). Otsu based optimal multilevel image thresholding using firefly algorithm. *Modelling and Simulation in Engineering*, 2014, 794574, 17.
18. Tsallis, C. (1988). Possible generalization of Boltzmann–Gibbs statistics. *Journal of Statistical Physics*, 52(1), 479–487.
19. Sadek, S. & Al-Hamadi, A. (2015). Entropic image segmentation: A fuzzy approach based on Tsallis entropy. *International Journal of Computer Vision*, 5(1), 1–7.
20. Sarkar, S., Paul, S., Burman, R., Das, S. & Chaudhuri, S.S. (2014). A fuzzy entropy based multi-level image thresholding using differential evolution. *Lecture Notes in Computer Science*, 8947, 386–395.
21. Anusuya, V. & Latha, P. (2014) A novel nature inspired Fuzzy Tsallis entropy segmentation of magnetic resonance images. *Neuroquantology*, 12(2), 221–229.
22. Kannappan, P.L. (1972). On Shannon's entropy, directed divergence and inaccuracy. *Probability Theory and Related Fields*, 22, 95–100.
23. Paul, S. & Bandyopadhyay, B. (2014). A novel approach for image compression based on multi-level image thresholding using Shannon entropy and differential evolution. In: *IEEE Students' Technology Symposium (TechSym)*, 56–61.
24. Kapur, J.N., Sahoo, P.K. & Wong, A.K.C. (1985). A new method for gray-level picture thresholding using the entropy of the histogram. *Computer Vision, Graphics, and Image Processing*, 29, 273–285.
25. Manic, K.S., Priya, R.K. & Rajinikanth, V. (2016). Image multithresholding based on Kapur/Tsallis entropy and firefly algorithm. *Indian Journal of Science and Technology*, 9(12), 89949.
26. Akay, B. (2013). A study on particle swarm optimization and artificial bee colony algorithms for multilevel thresholding. *Applied Soft Computing Journal*, 13(6), 3066–3091.
27. Dougherty, E.R.. (1994). *Digital Image Processing Methods*, 1st Edition, CRC Press.
28. Satapathy, S.C., Raja, N.S.M., Rajinikanth, V., Ashour, A.S. & Dey, N. (2018). Multi-level image thresholding using Otsu and chaotic bat algorithm. *Neural Computing and Applications*, 29(12), 1285–1307.
29. Rajinikanth, V. & Couceiro, M.S. (2015). RGB histogram based color image segmentation using firefly algorithm. *Procedia Computer Science*, 46, 1449–1457.
30. Hore, A. & Ziou, D. (2010). Image quality metrics: PSNR vs. SSIM. In: *IEEE International Conference on Pattern Recognition (ICPR)*, Istanbul, Turkey, 2366–2369.
31. Wang, Z., Bovik, A.C., Sheikh, H.R. & Simoncelli, E.P. (2004). Image quality assessment: From error measurement to structural similarity. *IEEE Trans Image Process*, 13(1), 1–14.
32. Lee, S.U., Chung, S.Y. & Park R.H. (1990). A comparative performance study techniques for segmentation. *Computer Vision Graphics and Image Processing*, 52(2), 171–190.

33. Sezgin, M. & Sankar, B. (2004). Survey over image thresholding techniques and quantitative performance evaluation. *Journal of Electronic Imaging* 13(1), 146–165.
34. Ghamisi, P., Couceiro, M.S., Benediktsson, J.A. & Ferreira, N.M.F. (2012). An efficient method for segmentation of images based on fractional calculus and natural selection. *Expert Systems with Applications*, 39(16), 12407–12417.
35. Eberhart, R.C. & Shi, Y. (2001). Tracking and optimizing dynamic systems with particle swarms. In *Proceedings of the IEEE congress on evolutionary computation (CEC)* (pp. 94–100), Seoul, Korea. Piscataway: IEEE.
36. Eberhart, R.C., Simpson, P.K. & Dobbins, R.W. (1996). *Computational Intelligence PC Tools*. Boston: Academic Press.
37. Blackwell, T.M. (2005). Particle swarms and population diversity. *Soft Computing*, 9, 793–802.
38. Clerc, M. (2006). *Particle Swarm Optimization*. London: ISTE.
39. Iwasaki, N. & Yasuda, K. (2005). Adaptive particle swarm optimization using velocity feedback. *International Journal of Innovative Computing, Information and Control*, 1(3), 369–380.
40. Passino, K.M. (2002). Biomimicry of bacterial foraging for distributed optimization and control. *IEEE Control Systems Magazine*, 22(3), 52–67.
41. Das, S., Biswas, A., Dasgupta, S. & Abraham, A. (2009) Bacterial foraging optimization algorithm: Theoretical foundations, analysis, and applications. In: Abraham, A., Hassanien, AE., Siarry, P., Engelbrecht, A. (eds) *Foundations of Computational Intelligence Volume 3. Studies in Computational Intelligence*, vol 203. Springer, Berlin, Heidelberg.
42. Liu, Y. & Passino, K.M. (2002). Biomimicry of social foraging bacteria for distributed optimization: Models, principles, and emergent behaviors. *Journal of Optimization Theory and Applications*, 115(3), 603–628.
43. Abraham, A., Biswas, A., Dasgupta, S. & Das, S. (2008). Anaysis of reproduction operator in bacterial foraging optimization. In: IEEE Congress on Evolutionary Computation CEC 2008. *IEEE World Congress on Computational Intelligence, WCCI 2008*, 1476–1483. IEEE Press, USA.
44. Yang, X.S. (2009). Firefly algorithms for multimodal optimization. In: *Proceeding of the Conference on Stochastic Algorithms: Foundations and Applications*, 169–178.
45. Yang, X.S. (2010). Firefly algorithm, levy flights and global optimization. In: Watanabe, O. & Zeugmann, T. (eds.) *Research and Development in Intelligent Systems XXVI*, 209–218. Springer, Berlin.
46. Yang, X.S. (2013). Multiobjective firefly algorithm for continuous optimization. *Engineering with Computers*, 29, 175–184
47. Yang, X.S. (2010). Firefly algorithm, stochastic test functions and design optimisation. The International *Journal of Bio-Inspired Computing*, 2(2), 78–84.
48. Yang, X.S. (2011). Review of meta-heuristics and generalised evolutionary walk algorithm. *The International Journal of Bio-Inspired Computing* 3(2), 77–84.
49. Dey, N. (2020). Applications of firefly algorithm and its variants. *Springer Tracts in Nature-Inspired Computing*, Springer.
50. Yang, X.-S. (2010). A new metaheuristic bat-inspired algorithm. In: González, J.R., Pelta, D.A., Cruz, C., Terrazas, G., & Krasnogor, N. (eds.) *NICSO 2010*. SCI, 284, 65–74. Springer, Heidelberg.
51. Yang X.-S. (2010). Nature-inspired metaheuristic algorithms. Luniver Press.
52. Dey, N. & Rajinikanth, V. (2020). Applications of bat algorithm and its variants. *Springer Tracts in Nature-Inspired Computing*, Springer.
53. Yang, X.-S. & Deb, S. (2009) Cuckoo search via Lévy flights. In: *Nature & biologically inspired computing, 2009*. NaBIC 2009. World Congress on IEEE.

54. Burnwal, S. & Deb, S. (2013). Scheduling optimization of flexible manufacturing system using cuckoo search-based approach. *The International Journal of Advanced Manufacturing Technology*, 64(5–8), 951–959.

55. Satapathy, S. & Naik, A. (2016). Social group optimization (SGO): A new population evolutionary optimization technique. *Complex & Intelligent Systems*. 2(3), 173–203.

56. Naik, A., Satapathy, S.C., Ashour, A.S. & Dey, N. (2016). Social group optimization for global optimization of multimodal functions and data clustering problems. *Neural Computing & Applications*, https://doi.org/10.1007/s00521-016-2686-9.

57. Rajinikanth, V. & Satapathy, S.C. (2018). Segmentation of ischemic stroke lesion in brain MRI based on social group optimization and fuzzy-Tsallis entropy. *The Arabian Journal for Science and Engineering*, 43, 4365–4378. https://doi.org/10.1007/s13369-017-3053-6

58. Dey, N., et al. (2019). Social-group-optimization based tumor evaluation tool for clinical brain MRI of Flair/diffusion-weighted modality.*Biocybernetics and Biomedical Engineering*, 39(3), 843–856.

59. Rao, R.V. (2016). *Teaching Learning-Based Optimization Algorithm*, Springer.

60. Rajinikanth, V., et al. (2017). Entropy based segmentation of tumor from brain MR images–A study with teaching learning based optimization. *Pattern Recognition Letters*, 94, 87–95.

61. Rao, R.V. (2016). *Jaya: An Advanced Optimization Algorithm and Its Engineering Applications*, Springer.

62. Rao, R.V. & Rai, D.P. (2017). Optimisation of welding processes using quasi-oppositional-based Jaya algorithm. *Journal of Experimental & Theoretical Artificial Intelligence*, 29(5), 1099–1117.

63. Rao, R.V., Saroj, A., Ocloń, P., Taler, J. & Taler, D. (2017). Single- and multi-objective design optimization of plate-fin heat exchangers using Jaya algorithm. *Heat Transfer Engineering*, 39(13-14), 1201–1216.

4 Image Segmentation

Medical image assessment based on specific computer algorithms is gaining popularity due to its accuracy and adaptability to a variety of images with different dimensions and orientations. During image assessment, the extraction and evaluation of the infected section using a suitable procedure is largely implemented to examine the greyscale and RGB-scale images [1–4].

Segmentation is one of the important image processing techniques implemented to extract a particular section of the image. A considerable number of the SOI extraction techniques are available in the literature. Based on the implementation, it can be classified as (i) automated segmentation and (ii) semi-automated segmentation techniques, the choice of which depends on the expertise of the operator and the complexity of the test image to be examined [5–7].

4.1 REQUIREMENT OF IMAGE SEGMENTATION

In medical image assessment, the segmentation implemented is normally used to extract the infected section from the clinical-grade test image for the evaluation and treatment planning process. The combination of multi-level thresholding and segmentation is executed by the researchers to examine a class of diseases using the images. These processes also help form a hybrid image processing procedure which offers better evaluation compared to traditional techniques. The features extracted from the segmented image section are considered to develop and implement a class of Machine Learning (ML) systems, the accuracy of which depends mainly on the segmented disease section. Further, the features of the segmented section can be considered to improve disease detection accuracy in the machine learning system [8–10].

The major requirements of image segmentation are listed below:

i. Extraction and assessment of the disease-infected section from the two-dimensional clinical images. Figure 4.1 depicts the hybrid image processing scheme implemented to extract and evaluate the skin melanoma section from the digital dermoscopy image. The task is to extract the melanoma section which is achieved using a threshold process to enhance the image and to extract the SOI with a segmentation procedure. After extracting the essential section from the image, its value is then compared to the ground-truth and, based on the attained quality values, the performance of the segmentation technique is validated.

ii. Implementing and improving disease detection accuracy of the ML system. During this stage, the image features are initially extracted with a technique

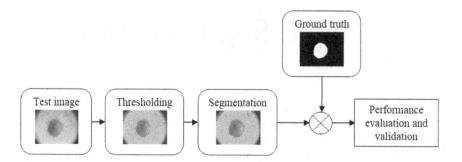

FIGURE 4.1 Hybrid Image Processing Technique Implemented to Examine the Skin Melanoma.

and, based on these features, classification systems are implemented and trained to support the automated detection of the disease.

iii.   Improving the performance of the traditional and modern DL system using the image features obtained from the segmented disease section.

4.2 EXTRACTION OF IMAGE REGIONS WITH SEGMENTATION

Extraction of the Section of Interest (SOI) from the medical image is the essential process which helps assess the disease in images with better diagnostic accuracy irrespective of the imaging modality. To achieve this, several segmentation procedures are proposed and implemented in the literature. This section presents a few commonly used image segmentation procedures employed in medical image processing.

4.2.1 MORPHOLOGICAL APPROACH

Morphological approach is used to pre-process images considered for the investigation through several operations implemented on the image pixels. This technique is widely utilized to smoothen the surface of the image. This procedure is a commonly used segmentation process, adopted to extract the particular pixel group based on what is needed.

Morphological enhancement and segmentation is one of the oldest image processing schemes. Due to its merit, morphology-based segmentation techniques are widely employed in a variety of segmentation procedures, one of which is the Markov Random Field segmentation (MRF) with Expectation Maximization (EM). The MRF-EM is an image segmentation technique employed to extract image regions into various sections. Information regarding the MRF-EM-based segmentation can be found in [11,12].

MRF-EM is a generally considered method for greyscale image segmentation problems. The necessary details are presented below:

Consider a greyscale trial picture $I = \{Y(M. N) | 0 \le Y \le L - 1$; in which Y symbolizes the strength of the picture at the pixel location (M, N) and L denotes the number of thresholds of the image (normally 256). Through segmentation, MRF will estimate the arrangement of every pixel by mapping them into a cluster of arbitrary labels defined as $X = \{x_1, ..., x_N\}|x_i \in l$. The number of labels is chosen as three in this work, which separates the test image into three sections.

Implementation of MRF-EM algorithm is defined below:

Step 1: Set the amount of labels (*l*) based on quantity of threshold values (*T*)
Step 2: Arrangement of cluster classes based on the selected *l*

$$k_1 = \{Y(M. N) = x_1 \,|0 \le Y \le t_1$$

$$k_2 = \{Y(M, N) = x_2 \,|t_1 \le Y \le t_2. \tag{4.1}$$

$$k_3 = \{Y(M, N) = x_3|t_2 \le Y \le L - 1$$

Step 3: Determine the initial parameter set $\Theta^{(0)}{}_i^l$ and likelihood probability function $p^{(0)}(f_i|x_1)$.
Step 4: Update the MRF model $x^{(t)}$ such that, the energy function U is minimised.

$$X^* = \underset{x}{argmin}\left\{\sum_i \left[\frac{(Y_i - \mu_{x_i})^2}{2\sigma_{x_i}^2} \ln \sigma_{x_i}\right] + \sum_{N_i} V_c(x)\right\} \tag{4.2}$$

where, N_i is a four pixel neighborhood and V_c is the clique potential.

Step 5: Implement the EM operation to update the parameter set $\Theta^{(i)}$ constantly till the log likelihood of $p^{(i)}(f|x)$ is maximized.
Step 6: Display the separated labels *as* the segmented results.

The performance of the MRF-EM algorithm is initially tested on greyscale and RGB-scale test images and then it is implemented on the images associated with the Salt & Pepper noise. Figure 4.2 depicts the brain MRI slice considered for the assessment and the aim is to segment the tumour section from the test image. This work employs a hybrid technique which integrates the morphology-based image enhancement using the Expectation Maximization (EM) operation and then employs a segmentation procedure, which separates the image section into three parts: normal brain section, tumor, and the background. The morphology and the EM help enhance the test image by executing a suitable morphological operation. This helps group the pixels to enhance the SOI to be extracted and evaluated.

Figure 4.2 (a) presents the test image. The MRF-EM segmentation is employed by fixing the iteration level at three. Initially, the EM algorithm is executed to improve the image and it creates the essential image labels (processed image) based on the EM value. The convergence of the EM operation is depicted in Figure 4.2 (b) and the enhanced image is presented in Figure 4.2 (c).

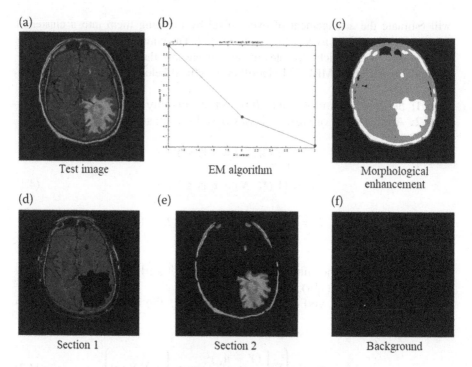

(a) Test image   (b) EM algorithm   (c) Morphological enhancement
(d) Section 1   (e) Section 2   (f) Background

FIGURE 4.2 Segmentation Results Attained for the Brain MRI Slice.

The segmentation results attained with the MRF-EM are illustrated in Figure 4.2 (d) to Figure 4.2 (f), where section 1 presents the normal brain section, section 2 depicts the tumor (SOI), and section 3 presents the background. After determining the tumor section, it is examined by a doctor to evaluate the severity of the disease.

The image is then tested against the medical image with the noise. The result is shown in Figure 4.3 (a) to (f). In this image, Figure 4.3 (b) is the initial morphological operation along with the noise. This operation is repeated to eliminate the noise pixels from the image to get the final enhancement image as in Figure 4.3 (c). The remaining results are similar to Figure 4.2. This confirms that the proposed procedure works well on the test image with/without noise, extracting the SOI accurately. The MRF-EM segmentation is also tested on the RGB-scale image (fundus retinal image), the results of which are depicted in Figure 4.4. This result also confirms that the MRF-EM approach works well on a class of test images, helping achieve better segmentation results to improve the disease diagnosis process. More information on the MRF-EM can be found in Rajinikanth et al. [13]. It is an automated segmentation technique existing to solve medical image analysis problems.

4.2.2 CIRCLE DETECTION

Hough Transform (HT) is a procedure meant to recognize the circles/circular-shaped objects from digital images [14–16].

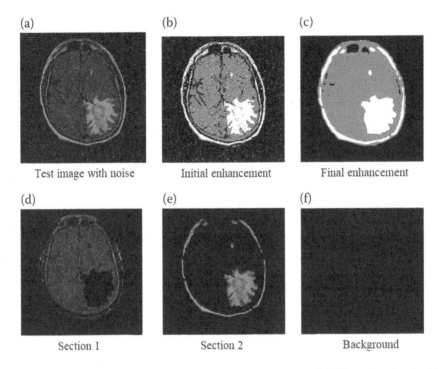

FIGURE 4.3 Segmentation Results Attained from the Brain MRI Slice Associated with Noise.

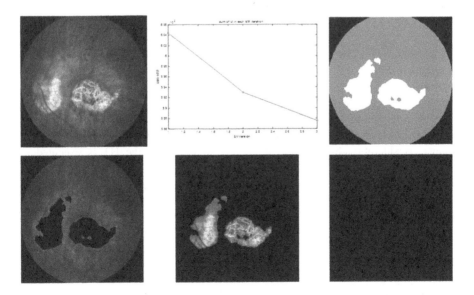

FIGURE 4.4 Segmentation of Abnormal Section from the RGB-Scale Retinal Image.

The HT employed in this work is discussed in [17].

Let, r = radius, X and Y = illustration axis, A and B = centre of a random circle. Then, it can be expressed as

$$(XA)^2 + (Y - B)^2 = r^2 \tag{4.3}$$

The HT is able to recognize the X and Y based on the preferred A, B, and r. The HT locates and traces the binary pixels (1s) in the test image after a possible border discovery procedure. The conventional HT traces and extracts all the existing extremely noticeable binary pixels (1s) in the digital illustration. In most cases, this HT fails to mine leukocyte segments from the hematological picture. HT-based circle detection procedure can be found in [18,19]. To increase the success rate in HT-based circle detection, the morphological procedure is included to eradicate the unnecessary but noticeable pixels in the pre-processed image.

The following steps present the modifications executed in HT to improve its segmentation accuracy:

Step 1: *Consider the pre-processed hematological illustration and execute border discovery.*
Step 2: *Implement the HT with a selected circle radius and recognize the entire pixel group with a magnitude unity.*
Step 3: *Extract the Haralick features and recognize the major-axis from the identified pixel groups.*
Step 4: *Implement the morphological action to eradicate the pixel groups, which is less than the identified major-axis.*
Step 5: *Improve the leukocyte segment with a pixel-level comparison.*
Step 6: *Extract the existing pixel collection and validate with the traces made by the HT.*

To demonstrate the HT-based circle detection process, the RGB-scale image depicted in Figure 4.5 (a) is considered. This image shows the availability of the leukocyte along with the blood stains. The primary objective is to extract the leukocyte with considerable accuracy. First, the test image is improved using a thresholding process, which will helps improve the visibility of the test image as depicted in Figure 4.5 (b). This step improves the stained leukocyte segment of the image considerably. Second, the segmentation based on the HT is executed to extract the leukocyte region as in Figure 4.5 (c) and (d). This step consists of procedures such as (i) detection of all the likely sections using HT, (ii) assessment of the section with large major-axis by means of the Haralick algorithm, (iii) implementing the morphological dilation and erosion of the pixel groups whose aspect is smaller than the major axis, and (iv) discovering and mining the binary form of the leukocyte. After extracting the leukocyte, the performance of the implemented HT is validated using a combative analysis between the extracted section and the ground-truth provided by an expert. Essential information on the HT-based circle detection can be found in [16–19].

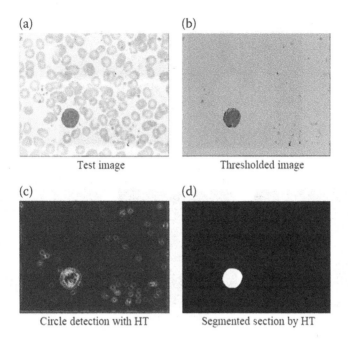

FIGURE 4.5 Detection and Extraction of Leukocyte Using HT.

4.2.3 Watershed Algorithm

Most illustrative processes include frailty assessment to extract irregular division from the chosen image. Recent works employed several semi/automated actions to extract the abnormal segments from an image. The Watershed-Algorithm (WSA) is a famous automated technique that helps to extract the SOI from the test image with accurately. The WSA implements the following processing operation: edge detection, watershed generation, morphological enhancement, and segmentation of the abnormal image section. Essential information regarding the WSA can be found in [20–22].

Figure 4.6 depicts the implementation of the WSA to extract the brain tumor from the MRI slice. Figure 4.6 (a) and (b) depicts the test image and the ground-truth (GT) image. Figure 4.6 (c) depicts the enhanced brain MRI using the thresholding process. Figure 4.6 (d) and (e) presents the results attained with the edge detection process and the watershed enhancement of the detected edge, respectively. Figure 4.6 (f) shows the enhanced tumour section based on the morphological enhancement and, finally, the extracted tumor is depicted in Figure 4.6 (g). A comparison between the extracted tumor and the GT is required to validate the performance of the implemented WSA.

4.2.4 Seed Region Growing

Region growing or the Seed Region Growing (SRG) is a semi-automated seg-mentation technique employed to identify and extract a particular pixel group. During this process, an operator initiates the seed point for the SRG and, as the

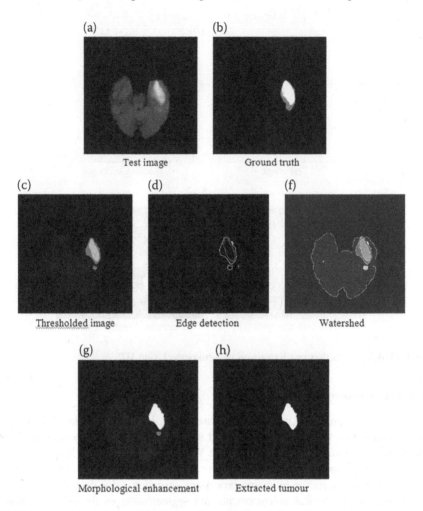

FIGURE 4.6 Watershed Algorithm-Based Extraction of the Tumor Section from Brain MRI Slice.

iteration increases, the seed continues to expand (grow) to identify all the similar connected pixels available in the image. The main limitation of this segmentation operation is the identification of a seed location in the image. When the operator starts the SRG a seed at a particular location, that region cultivates by attaching to similar neighboring pixels. This process is widely adopted by researchers to develop semi-automated image processing systems. The SRG approach works well on a class of images, such as the greyscale as well as RGB-scale images, and helps to segment the SOI accurately.

Figure 4.7 (a) and (b) depicts the implementation of the SRG on a chosen brain MRI slice. The task is to extract the visible section (stroke lesion) from the image. It is an RGB-scale image, which means the seed is initiated in an appropriate location more complex compared to greyscale images. When a seed is initiated by the operator, it will spread (grow) through the similar pixel group. This process stops after the maximum iteration has been reached. When the assigned iteration value is completed, the identified pixel group by the SRG is segmented and compared to the GT to confirm the performance of the segmentation technique. In the literature, a considerable number of procedures are utilized by SRG to extract a particular section, which can be found in [23-26].

4.2.5 PRINCIPAL COMPONENT ANALYSIS

PCA is one of the commonly accepted techniques in data analysis used in segmentation, feature selection, and classification. This approach initially sorts the information based on the need and identifies significant values to arrive at the necessary decision. It is a traditional data analysis procedure still widely used in image examination tasks. The application of the PCA in image segmentation is not significant. This section presents the results attained with the brain MRI slice with and without the noise section.

The PCA explores pixel values and helps extract the pixel group with the higher rank. The sample test image considered for the demonstration and the segmented binary form of the image is depicted in Figure 4.8. Figure 4.8 (a) presents the sample test image and Figure 4.8 (b) presents the extracted binary form of the tumour section.

4.2.6 LOCAL BINARY PATTERN

LBP a straightforward and efficient texture operator which labels each pixel of an illustration by thresholding the locality of every pixel and making the result a binary number. This procedure helps transform the actual image into a modified image in a binary of pixels (i.e., white section = 1 and black section = 0). This information can be easily processed using a computer algorithm which is employed to perform the essential task.

(a)   (b)

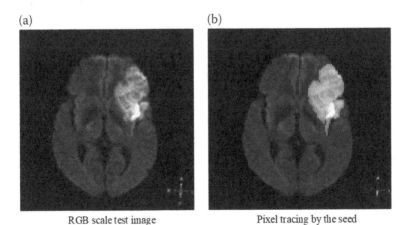

RGB scale test image   Pixel tracing by the seed

FIGURE 4.7 Extraction of the Stroke Lesion from the Brain MRI Slice Using the SRG Segmentation.

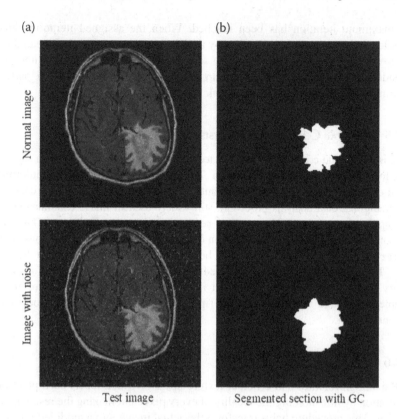

FIGURE 4.8   Extraction of the Brain MRI Slice Using the PCA Segmentation.

The concept behind the formation of the LBP operator is the two-dimensional exterior texture that can be described by two matching channels: local-spatial patterns and greyscale-contrast. The actual LBP operator creates labels for the pixels of a given image by thresholding the 3 x 3 locality of every pixel with respect to the middle value and outputting a binary numeral. The histogram of these 28 = 256 diverse labels can then considered a texture descriptor.

The subsequent information used for the LBP operator is illustrated by $LBP^S_{(M,N)}$, in which the subscript shown by means of the operator in a (M,N) locality and the superscript 'S' denotes the uniform patterns. All the remaining patterns are labeled with respect to a single label.

Finally, a new image is formed by the LBP with a description $F_{l(X,Y)}$ and a histogram value of

$$H_i = \sum_{X,Y} I\{F_{l(X,Y)} = i\}, \text{ for } i = 0, 1, 2, \ldots,n-1 \tag{4.4}$$

Other information on the LBP can be found in [27,28].

Figure 4.9 depicts the processing of a greyscale of a lung CT scan slice with the LBP segmentation technique. The selected test image and the segmented images are depicted in Figure 4.9 (a) and (b), respectively. After treating the test image using the LBP, the futures and the binary pixels can then be extracted, which helps classify/detect the disease from the image.

4.2.7 GRAPH CUT APPROACH

The Graph-Cut (GC) practice implements a graph theory to examine the image to attain rapid segmentation. It is a semi-automated process that helps create an operator-generated graph section on the illustration where every pixel is treated as a joint connected by weighted edges. When a graph is formed on the image, it recognizes all the pixels to be extracted and helps produce the binary form of the extracted section. Essential information on GC can be found in [29].

Figure 4.10 (a) and (b) depicts the result attained with the GC on the greyscale and RGB-scale images. This result depicts that the semi-automated GC helps extract the abnormality from the medical images with accurately. After extracting the section, it is then examined by the doctor during disease evaluation. The choice mainly depends on the operator and, though it is a semi-automated technique, it is largely adopted to extract a particular section in a class of digital images.

The GC works well on a class of greyscale and RGB-scale images. The only limitation is the physical identification of the required pixel values. The operator is responsible for identifying and marking the pixel groups to be extracted using the proposed segmentation.

The results attained with the dermoscopy image and lung CT scan slice are presented in Figure 4.10. This result confirms that the GC technique helps achieve better results on the images and helps obtain better segmentation results. The added merit of the GC is that it can extract essential pixel groups from an image associated with various noises.

(a)   (b)

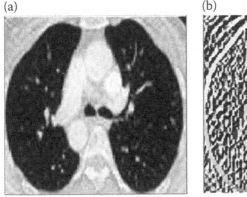
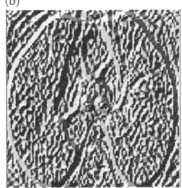

Lung CT scan slice   Enhanced image texture with LBP

FIGURE 4.9   Examination of a Lung CT Scan Slice with the LBP.

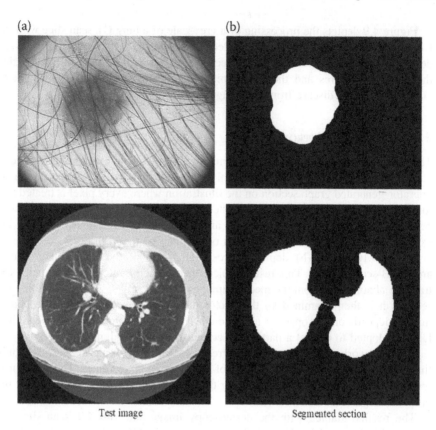

| Test image | Segmented section |

FIGURE 4.10   Segmentation Implemented with the Graph-Cut Technique.

4.2.8 CONTOUR-BASED APPROACH

Contour-assisted segmentation technique uses an adoptable structure, which helps identify and extract the essential pixel group according to what is required. The segmentation procedures, such as the Chan-Vese Segmentation (CVS), Active-Contour (AC), and Distance Regularised Level Set (DRLS), falls under the contour-based segmentation class. The operation of each approach is similar, in which an adaptable line is initiated as a box and it would adjust its orientation according to the pixel group where it is initiated and will try to collect the information from the images as the search iteration increases. The identified pixel groups are extracted and presented for further evaluation when the iteration level reaches a maximal value [29–32].

CVS is a common process extensively used to extort information from medical images. It works based on the energy minimization concept. In CVS, the curve is allowed to distinguish all the possible similar pixels existing in the picture. It is a semi-automated process, necessitating an operator to begin the elastic contour. The flexible contour segment regulates its direction when the iteration value rises until the matching pixel collection of the image is recognized. Finally, the exploration stops after reaching the minimal energy value and displays the district which lies surrounded by the curve.

Consider, Ω is a bordered position of $\mathfrak{R}^2$ with a border utility $\partial\Omega$. Assume: u_o: $\overline{\Omega} \rightarrow \mathfrak{R}$ is given image and C is the curvature segment.

If the section within C is Φ and exterior to the C is $\overline{\Omega}/\phi$.

If p1 and p2 denote the illustration pixels within and exterior to C, then the energy function will be E (e1, e2, C).

For the minimized energy, inf $_{e1, \ e2, \ C}$ E (e1, e2, C).

The working principle of other processes, such as the AC and DRLS, are similar to the CVS and this sub-section only presents the experimental results attained with the contour-based techniques.

Figure 4.11 presents the implementation of the CVS and the attained results. Initially, the image section to be extracted is assessed by the operator 'abs' based on the required pixel dimension. A bounding-box (initial contour) is to be implemented. The contour in the bounding-box is allowed to adjust its shape when the iteration increases and encircles the pixel groups to be extracted. After identifying all the essential pixels, the search process stops and the extracted section is presented as a binary scale image. The image section in Figure 4.11 clearly depicts the implementation of the CVS and the attained result.

Similar procedures are repeated using the DRLS and the AC and the attained results are depicted in Figures 4.12 (a) to (f) and 4.13 (a) to (d), respectively. All of the discussed contour-assisted methods are semi-automated segmentation techniques and will offer a better result when the process is appropriately implemented with a suitable bounding-box. The other applications of the contour-based methods can be found in [30,32].

4.2.9 CNN-Based Segmentation

Due to its merit, along with the image classification task, the Convolutional Neural Network (CNN) models are also used to segment the SOI from medical images. The following subsections present some of the recently developed CNN segmentation methods for discussion.

4.2.9.1 HRNet

HRNet was proposed in the Microsoft laboratory. It indicates enhanced presentation in the areas of semantic segmentation, illustration categorization, face detection, object recognition, and pose assessment. It considers the preparation of High Resolution (HR) images for demonstration. The active technique gets better demonstration of HR from representations of low resolutions formed by high to low resolution networks. In HRNet, the first stage HR network progressively enhances high to low resolution networks to organize more steps and connect the multi-resolution network in parallel [33].

HRNet is intelligent to support high-resolution depiction during procedures, as repeated multi-scale combinations are conducted by switching information through the multi-resolution parallel sub-networks continually during the process. The construction of the resulting network is displayed in Figure 4.14. This network has advantages in distinction to existing networks like SegNet, UNET, and Hourglass etc. These existing

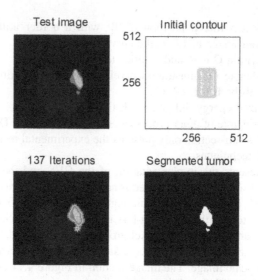

FIGURE 4.11   Segmentation of Tumour Section from Brain MRI with CVS.

networks lose a lot of necessary information in the progress of getting better HR from low-resolution representation. HRNet links high to low resolution networks in parallel instead of series, providing HR representation right through the procedure. Thus, the estimated heat map is more precise and spatially more accurate.

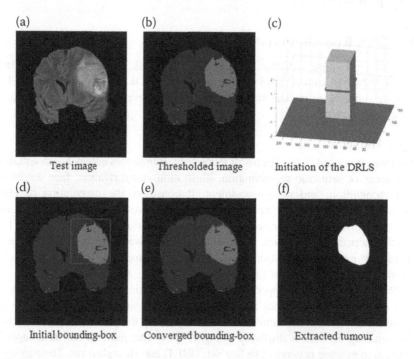

FIGURE 4.12   Segmentation of Tumour Section from Brain MRI Using DRLS.

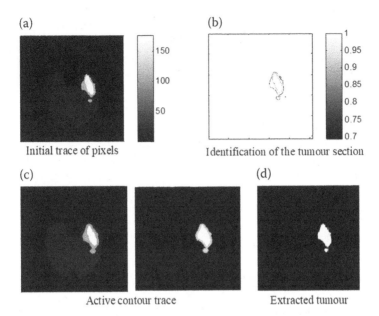

(a)

Initial trace of pixels

(b)

Identification of the tumour section

(c)

(d)

Active contour trace   Extracted tumour

FIGURE 4.13 Extraction of Tumour Section from Brain MRI Using AC.

Multi-Resolution Sequential Sub-Network

Existing models work by linking high to low resolution convolution sub-networks in series, where each individual sub-network forms a platform, a collection of an arrangement of convolutions. There is also a down sample layer through end-to-end sub-networks to split the resolution into halves.

Let $\mathcal{N}_{sr}$ be the subnet in the stage sth and resolution index r. First subnet resolution is given by $\frac{1}{2^{r}-1}$. The high-to-low system with S phases/stages can be indicated as:

$$\mathcal{N}_{11}\,\mathcal{N}_{22}\,\mathcal{N}_{33}\,\mathcal{N}_{44} \tag{4.5}$$

Multi-Resolution Parallel Sub-Network

Multi-resolution parallel sub-network starts the first phase/stage with a high-resolution subnet, slowly enhancing high to low resolution subnets, generating new phases/stages, and then associating the multi-resolution subnet in parallel. Eventually, the parallel subnet resolution of a later phase/stage comprises of the resolution from an earlier stage and below one stage. The network shown below contains 4 parallel subnets.

$$
\begin{array}{llll}
\mathcal{N}_{11} \searrow & \mathcal{N}_{21} \to & \mathcal{N}_{31} \to & \mathcal{N}_{41} \\
& \searrow \mathcal{N}_{22} \searrow & \mathcal{N}_{32} \to & \mathcal{N}_{42} \quad (4.6) \\
& & \searrow \mathcal{N}_{33} \searrow & \mathcal{N}_{43} \\
& & & \searrow \mathcal{N}_{44}
\end{array}
$$

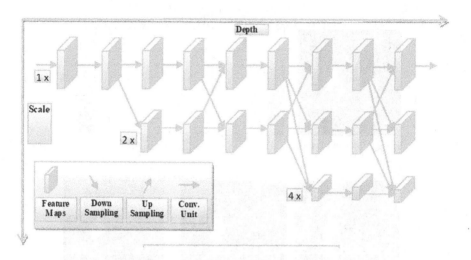

FIGURE 4.14   Conventional HRNet Architecture.

Multi-Scale Repeated Fusion

In this network, exchange units are introduced throughout the parallel subnet in such a way that an individual subnet continuously collects information from each one. The principle behind the exchange of information in this process is understood through an example. Here, the third stage is subdivided into multiple exchange blocks where every block consists of three parallel convolution modules, having exchange units followed by parallel units which is shown below.

$$
\begin{array}{ccccc}
C_{31}^{1} \searrow & & \nearrow C_{31}^{2} \searrow & & \nearrow C_{31}^{3} \searrow \\
C_{32}^{1} \longrightarrow & \varepsilon_{3}^{1} \longrightarrow C_{32}^{2} \longrightarrow & \varepsilon_{3}^{2} \longrightarrow C_{32}^{3} \longrightarrow & \varepsilon_{3}^{3} \quad (4.7) \\
C_{33}^{1} \nearrow & & \searrow C_{33}^{2} \nearrow & & \searrow C_{33}^{3} \nearrow
\end{array}
$$

where
   C_{sr}^{b} – convolution module,
   ε_{s}^{b} – exchange unit,
   and s is the stage, r is the resolution and b is the block.

Explanation of exchange units is illustrated by Figure 4.15. The input mapping is given by: $\{X_1, X_2, X_3, ...,X_s\}$ and the output mapping was given by: $\{Y_1, Y_2, Y_3, ...,Y_s\}$. The width and resolution of the output is the same as the input. Every output is a sum of input mapping i.e., $Y_K = \sum_{i=1}^{s} a(X_i, K)$. Assume 3×3 stride was done for down sampling and, for up sampling, 1×1 convolution (nearest neighbor).

HRNet experimental results (when tested with different datasets) show remarkable results for the applications such as face detection, semantic segmentation, and object detection.

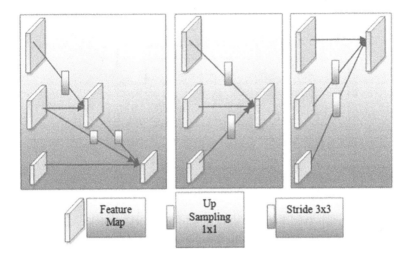

FIGURE 4.15 Information Exchange in HRNet.

4.2.9.2 SegNet

At the University of Cambridge, UK, a team from the robotics group researched and developed SegNet, which is a deep encoder/decoder architecture for multiclass pixel-wise segmentation [34]. The framework is comprised of an order of non-linear processing layers, called encoders, and a similar set of decoders. This comes after the last pixel-wise classifier. Generally, encoders are made up of a ReLU non-linearity and one or more convolutional layers with batch normalization, subsequently non-overlapping max-pooling and sub-sampling. Max-pooling indices are used in encoding sequences and, for up-sampling, sparse encoding is used especially in the pooling process of the decoder. The use of max-pooling indices in the decoders is an important feature of SegNet that executes the sampling of low resolution maps. One main advantage of SegNet is its ability to retain high frequency details for the segmented images. These images are enough to decrease the number of parameters the decoder needs for training. Using stochastic gradient descent, this framework can be trained end-to-end.

SegNet's architecture is shown in Figure 4.16. The encoder in SegNet is composed of 13 convolution layers which match with the 13 starting layers of VGG16, considered for classifying the objects.

Figure 4.17 illustrates the decoding method utilized by SegNet in which there is no learning implemented in the up-sampling stage. The up-sampling of the decoder network's feature map (input) is achieved by learned max-pooling indices from the equivalent encoder feature map. Dense feature maps are generated by combining feature maps and a trainable decoder channel.

SegNet is a deep network used for semantic segmentation. Basically, it was designed based on the principles of architecture for roads, outdoor, and indoor sites which is proficient together in terms of computational time and memory. The feature map's max-pooling indices are stored in SegNet and used in the decoder network for improved performance.

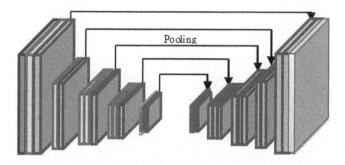

FIGURE 4.16 SegNet Architecture Used to Segment the RGB-Scale Image.

4.2.9.3 UNet

The UNet design is based on full convolution networks, adjusted such that it produces better segmentation results in medical imaging. UNet consists of two paths: contracting, and expansive. In the contracting path, it captures the context, whereas in expansive path, it enables exact localization. It includes two 3 x 3 convolutions, while the contracting path features the classical architecture of UNet that utilizes max-pooling operations with repetitive application. Figure 4.18 illustrates the architecture of UNet, a name given due to its 'U' shape. The main philosophy behind this network is it replaces pooling operations by using up-sampling operators. Ultimately, the resolution will increase layer by layer. This means the higher the number of layers, the higher the resolution of the image. Moreover, in every down-sampling, it doubles the feature channels.

Each stage in this expansive path involves up-sampling followed by (2 × 2) convolution that splits the number of feature channels into half. The contracting path crops the feature map because of the loss in border pixels in each convolution. The final layer is mapped by 1 × 1 convolutions that map all 64 units of feature vector. The network contains, in total, 23 convolutional layers. UNet performs well on image segmentation [35].

The cross-entropy loss function is united with the last feature map while training the UNet model and a pixel-wise Softmax is applied over it. The Softmax is denoted as:

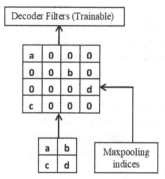

FIGURE 4.17 Decoding Process in SegNet.

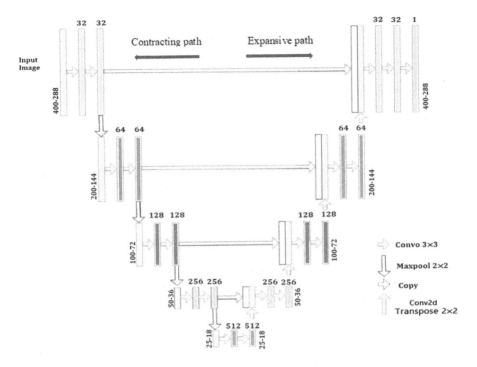

FIGURE 4.18   Conventional UNet Architecture.

$$p_k = \frac{e^{(a_k(x))}}{\sum_{k'=1}^{K} e^{(a_{k'}x)}} \tag{4.8}$$

In addition, the energy function is calculated by:

$$E = \sum_{x \in \Omega} w(x) \log(p_{l(x)}(x)) \tag{4.9}$$

where

  a_k: represents the activation in feature map k
  p_k: represents estimated maximum function
  K: no. of class
  $x \in \Omega$: pixel position
  $p_{l(x)}$: deviation

In the training data set, to counterbalance the diverse frequency of pixels from a specific class, the weight map is pre-calculated for ground-truth segmentation. The network is enforced to study the minor separation borders amid touching cells that are introduced.

The morphological operation used to calculate separation borders and the weight map is calculated using:

$$w(x) = w_c(x) + w_0 \cdot e^{\left(-\frac{(d_1(x)+d_2(x))^2}{2\sigma^2}\right)}$$
(4.10)

where

w : denotes the weight map

d_1: distance upto border of nearest first cell

d_2: distance upto border of nearest second cell

4.2.9.4 VGG UNet

Encoders-decoders, when combined, form the UNet architectures – which are famous for image segmentation in medical imaging and satellite images, etc. The weight of the pre-trained models, such as ImageNet, are used to initialize the weights of the neural network (i.e. trained on large dataset). This ensures better performance than other models, which are trained on small datasets from scratch. Model accuracy is very important in some applications like traffic safety and medicine. Pre-trained encoders enhance the architecture and performance of UNet. Applications such as object detection, image classification, and scene understanding have improved their performance after the introduction of CNN. Nowadays, CNN has outperformed human experts in several fields.

Image segmentation plays vital role in medical imaging as it enhances diagnostic capabilities. Fully Connected Network (FCN) is among the most popular state-of-the-art machine learning techniques. Segmentation accuracy attained by advancements in UNet consists of two paths: contracting and expansive. The contracting path sticks with the design of a convolutional network with pooling operations, alternating convolution, and gradually, down-sampling channels and expands feature map layers simultaneously. Each stage in the expansive path is composed of an up-sampling of the feature channel along with a convolution. The VGG Unet architecture is illustrated in Figure 4.19. The encoder for the UNet model is composed of 11 successive (series) layers in the VGG family, denoted by VGG-11. VGG-11 consists of seven convolution layers, each using rectified linear unit (ReLu) activation function, and five max-pooling operations, each reducing the feature channel by 2. The kernel size used for every convolutional layer is measured at 3 x 3 [36].

Common loss function, i.e., binary cross entropy can be used for classification problem where $\hat{y}_i$ denotes the prediction, y_i denotes the true value, and m denotes the number of samples:

$$H = -\frac{1}{m}\sum_{i=1}^{m}(y_i\log\hat{y}_i + (1-y_i)\log(1-\hat{y}_i))$$
(4.11)

To demonstrate the CNN segmentation procedure, the VGG UNet is considered and its experimental demonstration is implemented using the MATLAB software. Like other segmentation procedures, the VGG UNet also helps to attain a binary image as an outcome. The results attained in the present study for the brain tumor segmentation task is depicted in Figure 4.20. This confirms that the VGG UNet also

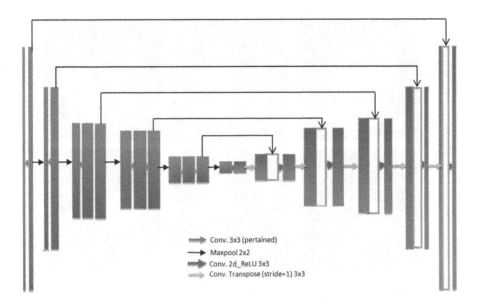

FIGURE 4.19   VGG UNet Structure to Support the Automated Segmentation.

presents the segmented result in binary image form and, based on this information, disease severity can be automatically examined.

4.3 ASSESSMENT AND VALIDATION OF SEGMENTATION

After executing a suitable image examination procedure to evaluate the disease from the medical images, its performance is validated to confirm clinical significance. A commonly implemented assessment procedure in the literature is a pixel-wise comparison between the ground-truth with the segmentation section. This comparison attains a number of performance measures and, based on the superiority of these measures, the significance of the segmentation procedure is therefore validated:

$$Jaccard\,Index\,(JI) = \frac{GTI \cap SI}{GTI \cup SI} = \frac{TP}{TP + FP + FN} \tag{4.12}$$

$$Dice\,(DI) = F1\,Score = \frac{2|GTI \cap SI|}{|GTI| + |SI|} = \frac{2TP}{2TP + FP + FN} \tag{4.13}$$

$$Accuracy\,(ACC) = \frac{TP + TN}{TP + TN + FP + FN} \tag{4.14}$$

$$Precision\,(PRE) = \frac{TP}{TP + FP} \tag{4.15}$$

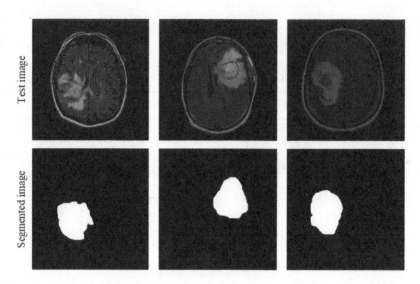

FIGURE 4.20   Segmentation Result Attained with VGG UNet for Brain MRI Slices.

$$Sensitivity\,(SEN) = \frac{TP}{TP + FN} \tag{4.16}$$

$$Specificity\,(SPE) = \frac{TN}{TN + FP} \tag{4.17}$$

$$Negative\ Predictive\ Value\,(NPV) = \frac{TN}{TN + FN} \tag{4.18}$$

where *GTI* is the ground-truth-image, *SI* is the segmented image, *TP*, *TN*, *FP* and *FN* denotes true-positive, true-negative, false-positive and false-negative, respectively.

During the assessment, the GTI and the SI pixels of similar dimension are compared and the essential values depicted are computed.

4.4 CONSTRUCTION OF CONFUSION MATRIX

Confusion Matrix (CM) is a graphical representation of the performance values attained during the validation of the segmentation outcome. To achieve this, the segmented image section is pixel-wise compared against the pixels of the GT and, based and the essential outcomes, are then recorded to evaluate the performance of the process.

The sample test images considered for the assessment is presented in Figure 4(a) to (c) and the segmentation result of the VGG UNet is then compared against the GT. This allows essential performance values to be computed. The CM constructed with the proposed technique is depicted in Figure 4.22. In the CM, the P and N represent the positive and negative pixels, respectively. The proposed comparison

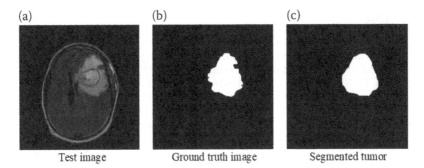

4.21 Comparison of VGG UNet Segmented Section with the GT Image.

Target value

TP=4320	FP=83	Sensitivity= 0.9081
FN=437	TN=60696	Specificity=0.9986
P=TP+FN=4757	N=FP+TN=60779	Accuracy=0.9921
Total pixels=P+N=65536		Precision=0.9811

Attained value

FIGURE 4.22 Confusion Matrix Used to Evaluate the Performance of the Segmentation Process.

helps to attain better values of JI (89.26%), DI (94.32%), and segmentation accuracy (99.21%). These measures are therefore sufficient to confirm the superiority of the proposed segmentation process. Thus, the VGG UNet offers better segmentation accuracy on the brain MRI segmentation task.

4.5 SUMMARY

This section presented various conventional and CNN-based segmentation techniques present in the literature to extract the SOI from medical images. In the hybrid image processing technique, segmentation plays a vital role. It extracts the disease-infected section with better accuracy. After the extraction, this section is then examined by the doctor personally or by using a computer software to evaluate the infection rate. The performance of the computer software responsible for disease assessment is validated with respect to the ground-truth to confirm its clinical significance. This section also discussed essential segmentation procedures with appropriate results attained using the MATLAB software.

REFERENCES

1. Priya, E., & Srinivasan, S. (2015). Automated identification of tuberculosis objects in digital images using neural network and neuro fuzzy inference systems. *Journal of Medical Imaging and Health Informatics*, 5, 506–512.
2. Priya, E., Srinivasan, S. & Ramakrishnan, S. (2014). Retrospective non-uniform illumination correction techniques in microscopic digital TB images. *Microscopy and Microanalysis*, 20(5), 1382–1391.
3. Akay, B. (2013). A study on particle swarm optimization and artificial bee colony algorithms for multilevel thresholding. *Applied Soft Computing Journal*, 13(6), 3066–3091.
4. Dougherty, E.R. (1994). *Digital Image Processing Methods*, 1st Edition, CRC Press.
5. Priya, E., & Srinivasan, S. (2016). Automated object and image level classification of TB images using support vector neural network classifier. *Biocybernetics and Biomedical Engineering*, 36(4), 670–678.
6. Priya, E. & Srinivasan, S. (2016). Validation of non-uniform illumination correction techniques in microscopic digital TB images using image sharpness measures. *The International Journal of Infectious Diseases*, 45(S1), 406.
7. Priya, E. & Srinivasan, S. (2017). Analysis of tuberculosis images using differential evolutionary extreme learning machines (DE-ELM). In: Hemanth, D.J. & Estrela, V.V. (eds) *Deep Learning for Image Processing Applications*, *Advances in parallel computing*, IOS Press, 111–136.
8. Bhandary, A., Prabhu, G.A., Rajinikanth, V., Thanaraj, K.P., Satapathy, S.C., Robbins, D.E., Shasky, C., Zhang, Y.D., Tavares, J.M.R.S. & Raja, N.S.M. (2020). *Pattern Recognition Letters*, 129, 271–278.
9. Dey, N., et al. (2019). Social-group-optimization based tumor evaluation tool for clinical brain MRI of Flair/diffusion-weighted modality. *Biocybernetics and Biomedical Engineering*, 39(3), 843–856.
10. Fernandes, S.L., Rajinikanth, V. & Kadry, S. (2019). A hybrid framework to evaluate breast abnormality using infrared thermal images. *The IEEE Consumer Electronics Magazine*, 8(5), 31–36.
11. Meyer, F. & Beucher, S. (1990). Morphological segmentation. *Journal of Visual Communication and Image Representation*, 1(1), 21–46.
12. Carson, C., Belongie, S., Greenspan, H. & Malik, J. (2002). Blobworld-image segmentation using expectation-maximization and its application to image querying. *IEEE Transactions on Pattern Analysis and Machine Intelligence*, 24(8), 1026–1038.
13. Rajinikanth, V., Raja, N.S.M. & Kamalanand, K. (2017). Firefly algorithm assisted segmentation of tumor from brain MRI using Tsallis function and Markov random field. *Journal of Control Engineering and Applied Informatics*, 19(3), 97–106.
14. Xu, L. & Oja, E. (1993). Randomized hough transform (RHT): basic mechanisms, algorithms, and computational complexities. *CVGIP: Image Understanding*, 57(2), 131–154. doi: 10.1006/ciun.1993.1009.
15. Xu, L., Oja, E. & Kultanen, K. (1990). A new curve detection method: Randomized Hough transform (RHT). *Pattern Recognition Letters*, 11(5), 331–338. doi: 10.1016/0167-8655(90)90042-Z.
16. Illingworth, J. & Kittler, J. (1988). A survey of the Hough transform. *Computer Vision, Graphics, and Image Processing*, 44(1), 87–116. doi: 10.1016/S0734-189X(88)80033-1.
17. Mukhopadhyay, P. & Chaudhuri, B.B. (2015). A survey of Hough transform. *Pattern Recognition*, 48(3), 993–1010.
18. Prinyakupt, J. & Pluempitiwiriyawej, C. (2015). Segmentation of white blood cells and comparison of cell morphology by linear and naïve Bayes classifiers. *BioMedical Engineering Online* 14, 63. doi: 10.1186/s12938-015-0037-1.

19. Cherabit, N., Chelali F.Z. & Djeradi, A. (2012). Circular hough transform for Iris localization. *Science and Technology*, 2(5), 114–121. doi: 10.5923/j.scit.20120205.02.

20. Rajinikanth, V., Satapathy, S.C., Dey, N. & Vijayarajan, R. (2018). DWT-PCA image fusion technique to improve segmentation accuracy in brain tumor analysis. *Lecture Notes in Electrical Engineering*, 471, 453–462.

21. Rajinikanth, V., Thanaraj, K.P., Satapathy, S.C., Fernandes, S.L. & Dey, N. (2019). Shannon's entropy and watershed algorithm based technique to inspect ischemic stroke wound. *Smart Innovation, Systems and Technologies*, 105, 23–31.

22. Shree, T.D.V., Revanth, K., Raja, N.S.M. & Rajinikanth, V. (2018). A hybrid image processing approach to examine abnormality in retinal optic disc. *Procedia Computer Science*. 125, 157–164.

23. Thanaraj, R.I.R., Anand, B., Rahul, A.J. & Rajinikanth, V. (2020). Appraisal of breast ultrasound image using Shannon's thresholding and level-set segmentation. *Advances in Intelligent Systems and Computing*, 1119, 621–630.

24. Raja, N.S.M., Rajinikanth, V., Fernandes, S.L. & Satapathy, S.C. (2017). Segmentation of breast thermal images using Kapur's entropy and hidden Markov random field. *Journal of Medical Imaging and Health Informatics*, 7(8), 1825–1829.

25. Nair, M.V., Gnanaprakasam, C.N., Rakshana, R., Keerthana, N. & Rajinikanth, V. (2018). *International Conference on Recent Trends in Advance Computing (ICRTAC)*, IEEE, 174–179.

26. Rajinikanth, V., Raja, N.S.M. & Arunmozhi, S. (2019). ABCD rule implementation for the skin melanoma assesment–A study. In: *IEEE International Conference on System, Computation, Automation and Networking (ICSCAN)*, 1–4.

27. Meena, K. & Suruliandi, A. (2011). Local binary patterns and its variants for face recognition. *International Conference on Recent Trends in Information Technology (ICRTIT)*, IEEE. doi: 10.1109/ICRTIT.2011.5972286.

28. Heikkilä, M., et al. (2009). Description of interest regions with local binary patterns. *Pattern Recognition*, 42(3), 425–436.

29. Boykov, Y. & Jolly, M.P. (2000). Interactive organ segmentation using graph cuts. In: Delp S.L., DiGoia A.M., Jaramaz B. (eds) *Medical Image Computing and Computer-Assisted Intervention – MICCAI 2000. MICCAI 2000. Lecture Notes in Computer Science*, 1935, 276–286. Springer, Berlin, Heidelberg.

30. Li, C., Xu, C., Gui, C. & Fox, M.D. (2010). Distance regularized level set evolution and its application to image segmentation. *IEEE Transactions on Image Processing*, 19(12), 3243–3254.

31. Palani, T.K., Parvathavarthini, B. & Chitra, K. (2016). Segmentation of brain regions by integrating meta heuristic multilevel threshold with Markov random field, *Current Medical Imaging Reviews*, 12(1), 4–12.

32. Rajinikanth, V., Dey, N., Kumar, R, Panneerselvam, J. & Raja, N.S.M. (2019). Fetal head periphery extraction from ultrasound image using Jaya algorithm and Chan-Vese segmentation. *Procedia Computer Science*, 152, 66–73.

33. https://towardsdatascience.com/overview-of-human-pose-estimation-neural-networks-hrnet-higherhrnet-architectures-and-faq-1954b2f8b249.

34. Badrinarayanan, V., et al. (2017). SegNet: A deep convolutional encoder-decoder architecture for image segmentation. *IEEE Transactions on Pattern Analysis and Machine Intelligence*, 39(12), 2481–2495.

35. Ronneberger, O., Fischer, P. & Brox, T. (2015). U-Net: Convolutional networks for biomedical image segmentation. arXiv:1505.04597 [cs.CV].

36. Chang, S.W. & Liao, S.W. (2019). KUnet: Microscopy image segmentation with deep Unet based convolutional networks. *IEEE International Conference on Systems, Man and Cybernetics (SMC)*. doi: 10.1109/SMC.2019.8914048.

5 Medical Image Processing with Hybrid Image Processing Method

Disease identification by automated and semi-automated diagnostic systems plays a major role during treatment planning and implementation. The computer software developed to detect diseases based on medical images acts as a support system during disease prognosis. Implementation of a suitable diagnosis system will considerably reduce the burden on doctors involved in the disease detection process.

In the literature, a number of machine learning (ML) techniques are proposed and implemented by researchers to detect the disease through medical images. The use of ML systems has considerably increased due to its simplicity and detection accuracy. Earlier research work confirms that the ML-based disease detection plays a vital role in detecting a variety of diseases with higher accuracy. The choice and implementation of a chosen ML technique depends on the availability of the computer software and the expertise of the scientist who developed the software. The ultimate aim of the ML scheme is to provide improved disease detection accuracy and to adapt for use on images recorded with varied modalities.

5.1 INTRODUCTION

The common procedures existing in the ML-based disease detection system involve (i) initial treatment of the medical images by resizing, removing noise, removing artifacts and providing initial corrections, (ii) pre-processing the image to enhance the disease section, (iii) post-processing the image to extract the disease-infected section through a segmentation process, (iv) feature extraction and its dimensionality reduction, and (v) classifier implementation and its performance evaluation.

Literature review reveals that the detection accuracy of an ML technique depends mainly on the classifier system, which works based on the extracted features. A two-class or a multi-class classification can be implemented according to what is necessary to detect illness in clinically obtained images. The brain tumor detection system is considered for this study.

Abnormalities in the brain affect the whole body, and any untreated illness may lead to problems ranging from mild disability to death [1–3]. The World Health Organization (WHO) has classified brain tumors into four categories based on the severity and rate of growth. Low-Grade Glioma (LGG), which is the affliction under consideration, is labeled as a Type-2 tumor [4,5].

The LGG (brain seizure) normally affects adults and is caused by abnormal growth in the brain of cells such as astrocytes and oligodendrocytes [6]. Early symptoms of LGG are invisible, only detectable imaging modalities such as Computed-Tomography (CT) or Magnetic-Resonance-Imaging (MRI). LGG is not as visible in CT/MRI, hence it requires a suitable procedure to recognize its size and orientation.

The early stages of LGG are usually diagnosed accidentally, when the patient is recommended a brain CT or MRI examination for some other illness. If the seizure is recognized in its early stages, then promising treatment procedures can be prescribed to manage and treat the affliction. Procedures such as radiation and chemotherapy, followed by surgery, are common existing recommendations to treat LGG. Severity may increase if it is diagnosed late. Due to its low visibility, detection of the LGG segment from the 3D/2D brain image needs a more efficient analysis. Earlier works on brain MRI assessment reveals that the integration of thresholding with segmentation helps attain better results. Further, the choice of thresholding and segmentation requires prior knowledge to attain the expected outcome. Previous literature reveal that the implementation of entropy-based techniques and between-class variance-based approaches can be implemented on some brain images [7–11]. Further, the implementation of semi/automated techniques is also discussed in detail with possible suggestions [3,12].

This proposal implements a CBP-based Otsu's threshold algorithm for initial identification of seizures followed by a segmentation procedure. The overall accuracy of CBP depends on the initial enhancement of the brain MRI based on an optimal threshold value. In this CBP technique, to minimize operator interaction, a soft-computing technique is implemented to compute the finest threshold that improves the visibility of the seizure.

The Brain-Storm-Optimization (BSO) approach proposed by Shi [34] is also considered, its convergence accuracy confirmed against the most successful methods such as Teaching-Learning-Based-Optimization (TLBO) and Social-Group-Optimization (SGO).

After possible improvements, the tumor fragment is then extracted using semi/automated segmentation techniques. Here, a complete evaluation of the existing segmentation techniques such as Watershed (WS), Level-Set (LS), Seed-Region-Growing (SRG), Active-Contour (AC), and Chan-Vese (CV) methods are presented, the choice of which is discussed with appropriate trials. Finally, a comparative assessment of the extracted tumor is made against the Synthetic Ground-Truth (SGT). The performance of the planned CBP is then established by calculating essential performance measure values, which substantiates the performance of CBP with a particular segmentation technique.

5.2 CONTEXT

Due to its clinical significance, brain abnormality assessments have recently been extensively discussed by researchers. This discussion varies from the enhancement of MRI, to the segmentation of abnormal section, to localization and feature extraction, classification with machine learning, and even deep-learning techniques.

The study by Maier et al. executed a forest-tree based system to segregate stroke sections from the brain MRI with improved performance [13]. Chaddad and Tanougast implemented a robust skull-stripping algorithm for brain MRIs. The authors confirmed that computerized techniques work well on the brain MRI without skull sector [14]. In their work, the axial view images were considered for evaluation.

The summary of a few of these MRI assessments are discussed in Table 5.1.

The proposed approach considered only the extraction of the abnormal section (seizure) from the 2D MRI slice with better performance. Further, this research also describes the choice of the axial slice compared to the others. Finally, the need for skull-stripping and the choice of appropriate segmentation technique are also discussed with relevant experiments.

5.3 METHODOLOGY

This research aims to implement a competent Computer-Based Practice (CBP) to assess the seizure section from the chosen MRI slice irrespective of its orientation. The overview of the methods considered in this work is illustrated in Figure 5.1. Primarily, the 3D reconstructed brain MRI for a subject under consideration is chosen for this study. The assessment of the 3D view is computationally complex, hence 2D slices are commonly used in radiology. In this research, the 2D slice from the 3D MRI is attained with ITK-Snap [25], which is a user-friendly tool that allows the user to view and extract the axial, coronal, and sagittal slices. This tool can also be used to generate the Synthetic Ground-Truth (SGT) for evaluation. The extracted 2D slice is then enhanced using a threshold system with Otsu's function and the BSO algorithm. This work considered a tri-level threshold to segregate the MRI slice into the background, normal brain, and seizure. After the segregation, the seizure is mined through segmentation. Finally, the performance of CBP is estimated by a comparative examination between the SGT and the extracted seizure.

5.3.1 DATABASE

The proposed work considered the LGG images of the Cancer Imaging Archive (TCGA), a large-scale, clinical-grade image database free to use for research purposes [26,27]. This dataset was contributed by Juratli et al. (2018), which holds the information of 243 patients, of which 135 are glioblastoma class, and 108 are LGG class) [28]. Other related research associated with the TCGA-LGG can be accessed from [29–31]. Every image of this dataset includes the skull section, making analysis complex. In the proposed work, 25 patients (25 dataset × 10 slices = 250 slices) were considered for the evaluation and the existing images are resized into 256 × 256 × 1 (greyscale) pixels to reduce the computation time of CBP.

Initial assessment is done by considering glioblastoma and LGG class 2D slices of a patient recorded with T2. This assessment considered the axial, coronal, and sagittal views which confirmed that the abnormal section is more

TABLE 5.1

Summary of Brain MRI Assessment Techniques

Reference	Methodology	Outcome
Acharya et al. [1]	Nonlinear feature extraction and classification.	This work proposed machine learning techniques to detect abnormalities in brain.
Fernandes et al. [2]	A frame work is proposed to evaluate the brain MRI.	The outcome of this technique confirmed efficiency based on a relative assessment.
Dey et al. [3]	This work implemented Social Group Optimization-based thresholding and segmentation techniques.	This technique implemented a procedure to evaluate the abnormal brain section using hybrid image examination procedures.
Satapathy and Rajinikanth [15]	Jaya algorithm-based threshlding and level set-based segmentation.	This work implemented hybrid segmentation procedures and the results authenticated that, their method works on a class of brain pictures.
Dey et al. [16]	GA-based Interval Filter for MRI de-noising.	This filter is employed using axial MRI slices where the proposed technique helps attain a noise-free MRI slice.
Jahmunah et al. [17]	Schizophrenia detection with the multi-channel EEG signal.	This work implemented an EEG analysis and mapped the analysis of the brain slice to obtain the correlation between the EEG and brain MRI.
Rajinikanth et al. [18]	Examination of brain tumor through Tsallis entropy and LS segmentation.	This work considered the benchmark dataset without the skull section and the result confirmed that the proposed work is a hybrid image processing tool.
Raja et al. [19]	Contrast enhance MRI is evaluated with Tsallis entropy and Level set segmentation.	The proposed work efficiently extracted the contrast-enhanced brain abnormality with greater accuracy.
Ma et al. [20]	Multi scale Active Contour for the MRI slice analysis.	This work helped extract multiple sections in brain MRI.
Tian et al. [21]	Morphology-based segmentation of fMRI.	This technique implemented a segmentation and region pixel-based vicinity-protecting projection on brain fMRI.
Chen et al. [22]	Level set-based MRI segmentation.	

(*Continued*)

TABLE 5.1 (Continued)

Reference	Methodology	Outcome
		This work effectively extracted the brain's caudate nucleus using the LS technique on MRI slices.
Elazab et al. [23]	A review of tumor modeling is presented.	This work discussed the available methods for tumor section modeling.
Menze et al. [24]	Assessment of various segmentation techniques.	This work released the BRATS database where the abnormal segment is analyzed with a segmentation approach and better results are attained.

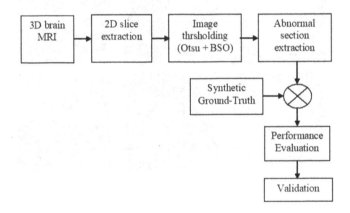

FIGURE 5.1   Various Stages Involved in the Proposed CBP.

visible in glioblastoma images compared to LGG. Further, all the views clearly showed the tumor section, but the intricacy in coronal- and sagittal-view images were higher compared to the axial-view due to the visibility of other brain tissues. Hence, in this research, only the axial view MRI slices are considered for assessment.

Figure 5.2 depicts the image classes such as glioblastoma (Figure 5.2 (a)) and LGG (Figure 5.2 (b)). With the help of ITK-Snap, the essential SGT is also extracted from these images to support the performance evaluation task. Figure 5.3 illustrates the result from the ITK-Snap, in which Figure 5.3 (a) depicts the trial image, Figure 5.3 (b) shows the generated binary mask and Figure 5.3 (c) presents the extracted SGT. The same is done on additional images (250 slices), and the generated SGT is then considered for the performance confirmation.

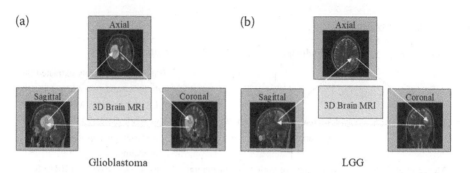

FIGURE 5.2   Comparison of Brain Abnormality Classes.

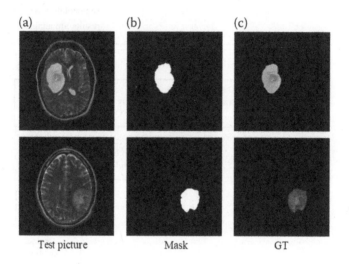

FIGURE 5.3   Development of synthetic Ground-Truth with ITK-Snap.

5.3.2 THRESHOLDING

Thresholding is a significant pre-processing technique considered to enhance test pictures generated by imaging modalities. In the literature, the bi-level and multi-level threshold is widely used in a variety of image processing tasks. In this research, the Otsu's approach is considered for the assessment.

5.3.3 OTSU'S FUNCTION

Otsu's threshold is usually considered to improve the test pictures in various fields [32,33]. Multi-thresholding is a broadly used scheme which separates the picture into different sections based on a selected threshold. Here, a 3-level threshold is applied to separate the 2D brain MRI into the background, normal part, and seizure. The Otsu's utility helps to develop the exterior of the seizure by dropping the regular brain sections. The traditional Otsu's technique is discussed in sub-section 3.4.1.

In the proposal, the BSO is employed to identify *Maximize J(Th)* in a hapha-zardly selected practice.

5.3.4 BRAIN STORM OPTIMIZATION

BSO was invented by Shi [34] for modeling the conclusion/discussion-making practice among humans in complex situations. Due to its performance, a con-siderable number of optimization tasks have been recently solved with the BSO [34–36]. This method works based on the problem-solving characteristic in-nate in humans. It is illustrated in the following steps:

Step 1: Create a brainstorming cluster with people of mixed backgrounds
Step 2: Begin the algorithm to create a selection of ideas
Step 3: Recognize and chose ideas to resolve the problem
Step 4: Choose the proposal with the highest likelihood
Step 5: Authorize the leading individual to collect several ideas with potential to resolve the specified mission
Step 6: Replicate the step until an improved resolution is attained for the selected problem

The final step produces an individual with the idea, and its model is presented in Equation (5.1):

$$X_{new}^D = X_{perfered}^D + \xi * n(\mu, \sigma) \tag{5.1}$$

where, X_{new}^D = final individual with an idea, $X_{perfered}^D$ = selected person, $n(\mu, \sigma)$ = Gaussian Random function with zero mean (μ) and a unity variance (σ).

$$\xi = \log sig ((0.5 * maximum\ iteration - current\ iteration)/K) \& rand () \tag{5.2}$$

In which k = slope of log sig ().

This research chose the traditional BSO existing in [34] to identify the finest threshold by maximizing the objective value. The primary BSO constraints are allocated as: number of agents = 25, dimension = 3, iteration = 2000 and stopping criteria = maximized Otsu's value. Other details on BSO can be found in [37]. The performance of the BSO is then compared against other related heuristic methods with similar as-signed constraints, such as Teaching Learning Based Optimization (TLBO) [38,39] and Social Group Optimization (SGO) [40,41]. The implementation of BSO is quite simple.

5.3.5 SEGMENTATION

Segmentation is considered to determine the abnormal section from the brain picture.

• **Watershed**

The infected fragment of LGG brain image may be extracted with the semi/automated techniques. The semi-automated exercise requires the help of a human-operator to start the task. Hence, automatic practice is widely preferred especially for medical purposes. This work implements the Marker-Controlled-Watershed-Mining (MCWS) to mine the irregular segment in the MRI. MCWS involves perimeter detection, formation of WS, and modifying the marker rate to mark the pixel groups, by augmenting the recognized pixels and determining the accepted region [42,43]. MCWS helps to extract all the probable pixels from the MRI. A comparable practice is then implemented for other images.

• **Level-Set**

This work implements the DRLS by Li et al. [44]. DRLS works based on the energy minimization course and it is articulated as:

$$\mathfrak{R}_e(\phi) \overset{\Delta}{=} \int_\Omega e(|\nabla\phi|) dX \tag{5.3}$$

where e = energy density value with $e = [0, \alpha] \to \mathfrak{R}$

In this research, an elastic curve is allowed to identify all the probable pixels related to the abnormal section active in the image. After discovering all these pixel groups, it will mine the area which is within the converged arc for assessment.

• **Seed-Region-Growing**

It is a semi-automated segmentation procedure in which an initiated seed grows rapidly to identify similar pixels, sharing like structures as the seeded section. The SRG works well on smooth surfaces but generates poor results if the seeded area has uneven pixel distribution. It is one of the classical segmentation techniques considered by researchers to extract the required section from the image under study. Related information on SRG can be found in [45–47].

• **Active-Contour**

Active contour is one of the most successful adjustable snake (line)-based technique normally considered to extract the chosen regions from medical images. In this work, the AC is initiated and allowed to expand on the image section by discovering the identical pixels. This resulted in negative yield for the multi-class pixel group due to uneven and poor images. The convergence rate of AC is large compared to the LS and CV segmentation techniques. Essential information on AC can be found in [48–51].

• **Chan-Vese**

In bio-image examination, segmentation plays a very important role since the general correctness of the image assessment system depends on the performance of this phase. Here, CVS discussed in section 4.2.8 is employed to remove the SC segment from the picture where threshold is applied.

5.3.6 PERFORMANCE EVALUATION AND VALIDATION

Qualified appraisal of the extracted section and GT is carried out in the pixel-level to appraise the advantage of the proposed system. The essential events, like Jaccard (JI), Dice (DC), accuracy (ACC), precision (PRE), sensitivity (SEN), and specificity (SPE) are calculated along with the true positive/negative and false positive/negative [52,53]. The mathematical model of these measures is presented below [54]:

$$Jaccard = SGT \cap S/SGT \cup S \qquad (5.4)$$

$$Dice = 2(SGT \cap S)/|SGT| \cup |S| \qquad (5.5)$$

$$Sensitivity = \frac{T_{+ve}}{T_{+ve} + F_{-ve}} \qquad (5.6)$$

$$Specificity = \frac{T_{-ve}}{T_{-ve} + F_{+ve}} \qquad (5.7)$$

$$Accuracy = \frac{T_{+ve} + T_{-ve}}{T_{+ve} + T_{-ve} + F_{+ve} + F_{-ve}} \qquad (5.8)$$

$$Precision = \frac{T_{+ve}}{T_{+ve} + F_{+ve}} \qquad (5.9)$$

where SGT = ground-truth, S = segmented picture, T_{+ve}, T_{-ve}, F_{+ve} and F_{-ve} denotes true-positive, true-negative, false-positive and false-negative, respectively.

5.4 RESULTS AND DISCUSSION

This part of this research presents and discusses the outcomes reached with CBP. This research is undertaken with the MATLAB software. All the considered axial-view images were tested with the developed CBP and the average values of the performance measures were considered to authenticate the performance.

Initially, Otsu's thresholding process with BSO, TLBO and SGO were performed on the axial view of 2D slices and the enhanced image was then considered for evaluation. Figure 5.4 presents the convergence of the heuristic search for a preferred image with a tri-level threshold. This image confirms that, irrespective of the algorithm, the optimal threshold as well as the attained Otsu's function are approximately equal. Only the convergence time of the algorithm varies based on its complexity, which is not a major concern in the biomedical image evaluation task,

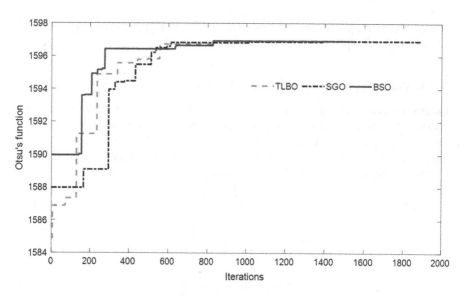

FIGURE 5.4   Convergence of Heuristic Search for the Tri-Level Optimization Problem.

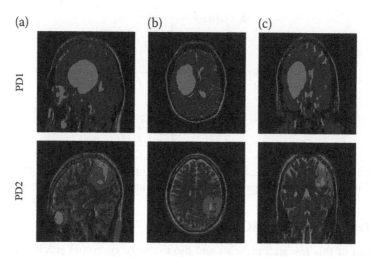

FIGURE 5.5   Outcome of Otsu's + BSO Thresholding. (a) Sagittal, (b) Axial, and (c) Coronal Views.

simply to achieve a better throughput during the image analysis task. In the proposed work, the BSO algorithm showed better convergence compared to alternative algorithms hence, for a mass screening task, this algorithm will provide better outcome.

The thresholding results attained with glioblastoma and LGG are presented in Figure 5.5. Figure 5.5 (a) to (c) presented the sagittal, axial, and coronal views

FIGURE 5.6 Results Attained with the Watershed Segmentation.

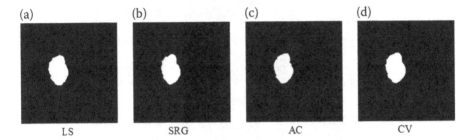

FIGURE 5.7 Results Attained with Other Segmentation Techniques.

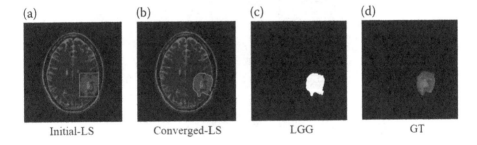

FIGURE 5.8 Results attained with the LS method on LGG sample image.

respectively. The tumor section enhanced in glioblastoma is more visible compared to the LGG. Hence, it is essential to choose an appropriate segmentation procedure, which will extract the tumor section with enhanced accuracy. If the 2D slices are free of skull, it is possible to use automated techniques like the watershed. Else, semi-automated techniques such as LS, SRG, AC and CV are preferred. Figure 5.6 presents the result attained for the glioblastoma images with the WS approach. The other results attained with LS, SRG, AC and CV are also depicted in Figure 5.7. This confirms that the proposed CBP works well on the 2D axial slice with the skull section. Thus, skull elimination is not essential in enhancing performance during the brain MRI assessment.

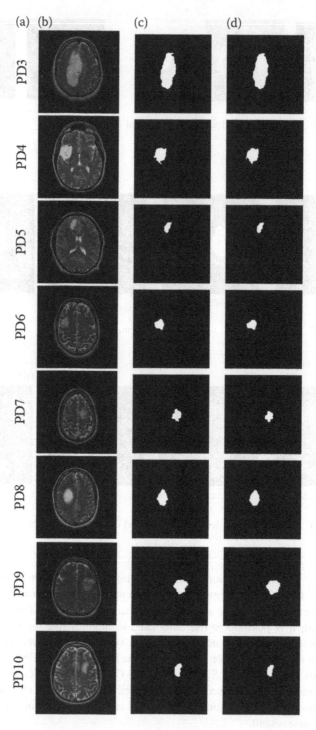

FIGURE 5.9 Results Attained with the Sample Test Images.

TABLE 5.2
Performance Measures Obtained for the Chosen LGG Images with LS Segmentation

Data	TPR	FNR	TNR	FPR	JI (%)	DI (%)	SEN (%)	SPE (%)	ACC (%)	PRE (%)
PD1	0.9706	0.0294	0.9993	0.0007	95.73	97.82	97.06	99.93	99.80	98.59
PD2	0.9108	0.0892	0.9986	0.0014	87.86	93.54	91.08	99.86	99.53	96.14
PD3	0.9629	0.0371	0.9962	0.0038	91.31	95.46	96.29	99.62	99.40	94.65
PD4	0.9434	0.0566	0.9988	0.0012	89.49	94.46	94.34	99.88	99.75	94.57
PD5	0.9704	0.0296	0.9993	0.0007	89.63	94.53	97.04	99.93	99.91	92.15
PD6	0.9475	0.0525	0.9986	0.0014	84.75	91.75	94.75	99.86	99.81	88.92
PD7	0.9411	0.0589	0.9993	0.0007	87.53	93.35	94.11	99.93	99.87	92.60
PD8	0.9843	0.0157	0.9992	0.0008	94.74	97.30	98.43	99.92	99.90	96.19
PD9	0.9501	0.0499	0.9986	0.0014	89.91	94.68	95.01	99.86	99.75	94.36
PD10	0.9941	0.0059	0.9978	0.0022	81.04	89.53	99.41	99.78	99.78	81.44

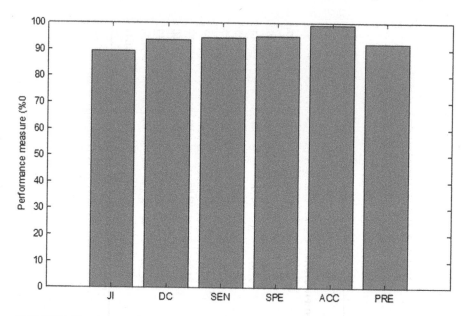

FIGURE 5.10   Average Performance Measures Obtained with the LGG Images.

A similar procedure is executed on the LGG images. Results confirm that only LS-based semi-automated technique offered better results compared to other segmentation techniques. These results are presented in Figure 5.8. With this, it is confirmed that the LS-based segmentation technique works well on the 2D slices of LGG in spite of the existence of the skull section.

The other LGG images (250 2D slices) considered in the study also undergo the same procedure. The sample results attained with the Otsu + BSO thresholding and LS segmentation is depicted in Figure 5.9. Figure 5.9 (a) presents the pseudo name and Figure 5.9 (b) to (d) depicts the test picture, binary SGT, and the extracted seizure, respectively. The performance measures attained for these images are depicted in Table 5.2. The values in this table confirm that the performance measures such as JI, DC, SEN, SPE, ACC, and PRE obtained in the present research using Otsu + BSO and LS are positive.

Better results were obtained when the procedure was implemented on the other 250 image slices. Figure 5.10 confirms that the average performance attained with the proposed CBP with LS on the LGG is good. Further, this technique provided an average segmentation accuracy of >99%.

The future scope of the present research is as follows: (i) Other patients' images of the TCGA-LGG could be examined with traditional and soft-computing based techniques, (ii) a skull-stripping method could be implemented to remove the skull section, allowing a suitable semi/automated segmentation technique to be implemented, and (iii) other segmentation procedures, such as clustering methods, local binary pattern, and principal component-based methods could be implemented for which its performance could be evaluated.

5.5 SUMMARY

This research proposed a Computer-Based Practice (CBP) to evaluate the seizure found in the brain MRI. During analysis, 2D axial-view MRI slices recorded with T2 are considered for assessment. Otsu and BSO-based thresholding were implemented to enhance the test picture. Later a specific segmentation technique is then implemented to extract the seizure, which is then compared to the Synthetic Ground-Truth obtained with ITK-Snap. The clinical-grade TCGA-LGG database was considered for the analysis and a sum of 250 images of dimensions 256 × 256 × 1 pixels were used for the evaluation. This work presented a detailed study on various segmentation techniques existing in the literature. The outcome of this research confirmed that the LS-based approach offered a segmentation accuracy of >99%. Thus, the proposed technique is clinically noteworthy and, in the future, it can also be considered to evaluate clinical-grade brain MRI slices.

REFERENCES

1. Acharya, U.R., et al. (2019). Automated detection of Alzheimer's disease using brain MRI images–a study with various feature extraction techniques. *Journal of Medical Systems*, 43(9), 302. doi: 10.1007/s10916-019-1428-9.
2. Fernandes, S.L., Tanik, U.J., Rajinikanth, V. & Karthik, K.A. (2019). A reliable framework for accurate brain image examination and treatment planning based on early diagnosis support for clinicians. *Neural Computing and Applications*, 1–12. doi: 10.1007/s00521-019-04369-5.
3. Dey, N., et al. (2019). Social-group-optimization based tumor evaluation tool for clinical brain MRI of Flair/diffusion-weighted modality. *Biocybernetics and Biomedical Engineering*, 39(3), 843–856. doi: 10.1016/j.bbe.2019.07.005.
4. Louis, D.N., et al. (2016). The 2016 World Health Organization classification of tumors of the central nervous system: a summary. *Acta Neuropathologica*. doi: 10.1007/s00401-016-1545-1.
5. Wesseling, P. & Capper, D. (2018). WHO 2016 classification of gliomas. *Neuropathology and Applied Neurobiology*, 44(2), 139–150. doi: 10.1111/nan.12432.
6. Forst, D.A., et al. (2014). Low-grade gliomas. *Oncologist*, 19(4), 403–413. doi: 10.1634/theoncologist.2013-0345.
7. Shree, N.V. & Kumar, T.N.R. (2018). Identification and classification of brain tumor MRI images with feature extraction using DWT and probabilistic neural network. *Brain Informatics*, 5, 23–30. doi: 10.1007/s40708-017-0075-5.
8. Tian, Z., Dey, N., Ashour, A.S., et al. (2017). Morphological segmenting and neighborhood pixel-based locality preserving projection on brain fMRI dataset for semantic feature extraction: an affective computing study. *Neural Computing & Applications*, 30(12), 3733–3748.
9. Kanmani, P. & Marikkannu, P. (2018). MRI brain images classification: a multi-level threshold based region optimization technique. *Journal of Medical Systems*, 42(4), 62.
10. Rajinikanth, V., Raja, N.S.M. & Kamalanand, K. (2017). Firefly algorithm assisted segmentation of tumor from brain MRI using Tsallis function and Markov random field. *Journal of Control Engineering and Applied Informatics*, 19(3), 97–106.
11. Amin, J., Sharif, M., Yasmin, M., et al. (2018). Big data analysis for brain tumor detection: deep convolutional neural networks. *Future Generation Computer Systems*, 87, 290–297.

12. Rajinikanth, V., Dey, N., Satapathy, S.C., et al. (2018). An approach to examine magnetic resonance angiography based on Tsallis entropy and deformable snake model. *Future Generation Computer Systems*, 85, 160–172.

13. Maier, O., Wilms, M., Gablentz, V.D.J., et al. (2015). Extra tree forests for sub-acute ischemic stroke lesion segmentation in MR sequences. *Journal of Neuroscience Methods*, 240, 89–100.

14. Chaddad, A. & Tanougast, C. (2016). Quantitative evaluation of robust skull stripping and tumour detection applied to axial MR images. *Brain Information*, 3(1), 53–61.

15. Satapathy, S.C. & Rajinikanth, V. (2018). Jaya algorithm guided procedure to segment tumor from brain MRI. *Journal on Optimization*, 2018, 3738049. doi: 10.1155/2018/3738049.

16. Dey, N., et al. (2015). Parameter optimization for local polynomial approximation based intersection confidence interval filter using genetic algorithm: An application for brain MRI image de-noising. *Journal of Imaging*, 1(1), 60–84.

17. Jahmunah, V., et al. (2019). Automated detection of schizophrenia using nonlinear signal processing methods. *Artificial Intelligence in Medicine*, 100, 101698. doi: 10.1016/j.artmed.2019.07.006.

18. Rajinikanth, V., Fernandes, S.L., Bhushan, B. & Sunder, N.R. (2018). Segmentation and analysis of brain tumor using Tsallis entropy and regularised level set. *Lecture Notes in Electrical Engineering*, 434, 313–321.

19. Raja, N.S.M., Fernandes, S.L. Dey, N., Satapathy, S.C. & Rajinikanth, V. (2018). Contrast enhanced medical MRI evaluation using Tsallis entropy and region growing segmentation. *Journal of Ambient Intelligence and Humanized Computing*, 1–12. doi: 10.1007/s12652-018-0854-8.

20. Nair, M.V., Gnanaprakasam, C.N., Rakshana, R., Keerthana, N. & V Rajinikanth, V. (2018). *International Conference on Recent Trends in Advance Computing (ICRTAC)*, IEEE, 174–179.

21. Maier, O., Wilms, M., Von der Gablentz, J., Krämer, U.M., Münte, T.F. & Handels, H. (2015). Extra tree forests for sub-acute ischemic stroke lesion segmentation in MR sequences. *Journal of Neuroscience Methods*, 240, 89–100. doi: 10.1016/j.jneumeth.2014.11.011.

22. Chen, Y., et al. (2019). A distance regularized level-set evolution model based MRI dataset segmentation of brain's caudate nucleus. *IEEE Access*, 7, 124128–124140.

23. Elazab, A., Abdulazeem, Y.M., Anter, A.M., et al. (2018). Macroscopic cerebral tumor growth modelling from multimodal images: a review. *IEEE Access*. doi: 10.1109/access.2018.2839681.

24. Menze, B.H., Jakab, A., Bauer, S., et al. (2015). The multimodal brain tumor image segmentation benchmark (BRATS). *IEEE Transactions on Medical Imaging*, 34(10), 1993–2024.

25. Yushkevich, P.A., Gao, Y. & Gerig, G. (2016). ITK-SNAP: An interactive tool for semi-automatic segmentation of multi-modality biomedical images. In: *38th Annual International Conference of the IEEE Engineering in Medicine and Biology Society (EMBC)*, IEEE, 3342–3345. doi: 10.1109/EMBC.2016.7591443.

26. http://cancergenome.nih.gov/.

27. Pedano, N., Flanders, A.E., Scarpace, L., Mikkelsen, T., Eschbacher, J.M., Hermes, B., et al. (2016). Radiology data from the cancer genome atlas low grade glioma [TCGA-LGG] collection. *The Cancer Imaging Archive*. doi: 10.7937/K9/TCIA.2016.L4LTD3TK.

28. Clark, K., Vendt, B., Smith, K., Freymann, J., Kirby, J., Koppel, P., Moore, S., Phillips, S., Maffitt, D., Pringle, M., Tarbox, L. & Prior, F. (2013). The Cancer Imaging Archive (TCIA): Maintaining and operating a public information repository. *Journal of Digital Imaging*, 26(6), 1045–1057.

29. Juratli, T.A., et al. (2018). Radiographic assessment of contrast enhancement and T2/FLAIR mismatch sign in lower grade gliomas: correlation with molecular groups. *Journal of Neuro-Oncology*, 2018. doi: 10.1007/s11060-018-03034-6.

30. Halani, S.H., Yousefi, S., Vega, J.V., Rossi, M.R., Zhao, Z., Amrollahi, F., Holder, C.A., Baxter-Stoltzfus, A., Eschbacher, J., Griffith, B., Olson, J.J., Jiang, T., Yates, J.R., Eberhart, C.G., Poisson, L.M., Cooper, L.A.D. & Brat, D.J. (2018). Multi-faceted computational assessment of risk and progression in oligodendroglioma implicates NOTCH and PI3K pathways. *Precision Oncology*, doi: 10.1038/s41698-018-0067-9.

31. Liu, Z., Zhang, T., Jiang, H., Xu, W. & Zhang, J. (2018). Conventional MR-based preoperative nomograms for prediction of IDH/1p19q subtype in low-grade glioma. *Academic Radiology*. doi: 10.1016/j.acra.2018.09.022.

32. Otsu, N. (1979). A threshold selection method from gray-level histograms. *Transactions on Systems, Man, and Cybernetics*, 9(1), 62–66.

33. Dey, N., Chaki, J., Moraru, L., Fong, S. & Yang, X.S. (2020). Firefly algorithm and its variants in digital image processing: A comprehensive review. In: *Applications of Firefly Algorithm and Its Variants*, 1–28, Springer, Singapore.

34. Shi, Y. (2011). Brain storm optimization algorithm. In: Tan, Y., Shi, Y., Chai, Y., Wang, G. (eds.) ICSI 2011, Part I, *LNCS*, vol. 6728, 303–309, Springer, Heidelberg.

35. Arsuaga-Ríos, M. & Vega-Rodríguez, M.A. (2014). Cost optimization based on brain storming for grid scheduling. In: *Proceedings of the 2014 Fourth International Conference on Innovative Computing Technology (INTECH)*, 31–36. Luton, UK.

36. Chen, J., Cheng, S., Chen, Y., Xie, Y. & Shi, Y. (2015). Enhanced brain storm optimization algorithm for wireless sensor networks deployment. In: *Proceedings of 6th International Conference on Swarm Intelligence (ICSI 2015)*, 373–381. Springer International Publishing, Beijing, China.

37. Cheng, S., Sun, Y., Chen, J., Qin, Q., Chu, X., Lei, X. & Shi, Y. (2017). A comprehensive survey of brain storm optimization algorithms. In: *Proceedings of 2017 IEEE Congress on Evolutionary Computation (CEC 2017)*, 1637–1644. IEEE, Donostia, San Sebastián, Spain.

38. Rao, R.V., Savsani, V.J. & Vakharia, D.P. (2012). Teaching-learning-based optimization: A novel optimization method for continuous non-linear large scale problems. *Information Sciences*, 183(1), 1–15.

39. Rao, R.V., Savsani, V.J. & Vakharia, D.P. (2011). Teaching-learning-based optimization: A novel method for constrained mechanical design optimization problems. *Computer-Aided Design*, 43(3), 303–315.

40. Satapathy, S. & Naik, A. (2016). Social group optimization (SGO): a new population evolutionary optimization technique. *Complex & Intelligent Systems*, 2(3), 173–203.

41. Dey, N., Rajinikanth, V., Ashour, A.S. & Tavares, J.M.R.S. (2018). Social group optimization supported segmentation and evaluation of skin melanoma images. *Symmetry* 10(2), 51. doi: 10.3390/sym10020051.

42. Roerdink, J.B.T.M. & Meijster, A. (2001). The watershed transform: definitions, algorithms and parallelization strategies. *Fundamenta Informaticae*, 41, 187–228.

43. Kaleem, M., Sanaullah, M., Hussain, M.A., Jaffar, M.A. & Choi, T.-S. (2012). Segmentation of brain tumor tissue using marker controlled watershed transform method. *Communications in Computer and Information Science*, 281, 222–227.

44. Li, C., Xu, C., Gui, C. & Fox, M.D. (2010). Distance regularized level set evolution and its application to image segmentation. *IEEE Transactions on Image Processing*, 19(12), 3243–3254.

45. Dehdasht-Heydari, R. & Gholami, S. (2019). Automatic Seeded Region Growing (ASRG) Using Genetic Algorithm for Brain MRI Segmentation. *Wireless Personal Communications*, 109(2), 897–908. doi: 10.1007/s11277-019-06596-4.

46. Shih, F. Y. & Cheng, S. (2005). Automatic seeded region growing for color image segmentation. *Image and Vision Computing*, 23(10), 877–886.
47. Malek, A.A., et al. (2010). Region and boundary segmentation of microcalcifications using seed-based region growing and mathematical morphology. *Procedia – Social and Behavioral Sciences*, 8, 634–639.
48. Yang, X. & Jiang, X. (2020). A hybrid active contour model based on new edge-stop functions for image segmentation. *International Journal of Ambient Computing and Intelligence (IJACI)*, 11(1), 87–98. doi: 10.4018/IJACI.2020010105.
49. Bresson, X., Esedoḡlu, S., Vandergheynst, P., Thiran, J.-P. & Osher, S. (2007). Fast global minimization of the active contour/snake model. *Journal of Mathematical Imaging and Vision*, 28(2), 151–167.
50. Chan, T.F. & Vese, L.A. (2002). Active contour and segmentation models using geometric PDE's for medical imaging. *Geometric Methods in Bio-medical Image Processing*, 63–75. doi: 10.1007/978-3-642-55987-7_4.
51. Chan, T.F. & Vese, L.A. (2001). Active contours without edges. *IEEE Transactions on Image Processing*, 10(2), 266–277.
52. Wang, R. & Wang, G. (2019). Web text categorization based on statistical merging algorithm in big data environment. *International Journal of Ambient Computing and Intelligence (IJACI)*, 10(3), 17–32. doi: 10.4018/IJACI.2019070102.
53. Ali, et al. (2019). Adam deep learning with SOM for human sentiment classification, *International Journal of Ambient Computing and Intelligence (IJACI)*, 10(3), 92–116. doi: 10.4018/IJACI.2019070106.
54. Yan, J., Chen, S. & Deng, S. (2019). A EEG-based emotion recognition model with rhythm and time characteristics. *Brain Informatics*, 6, 7. doi: https://doi.org/10.1186/s40708-019-0100-y.

6 Deep Learning for Medical Image Processing

The availability of high-end computing facilities and the necessary software allows the implementation of Deep-Learning Architectures (DLA) for the automated detection of diseases with higher accuracy, which is supported by the literature using examples of traditional and customary DLA [1–3].

The main advantage of the DLA is it is designed to examine RGB-scale images while also working well on a class of greyscale images. The main limitation of the DLA, on the other hand, is that before executing the detection process, the images to be examined requires resizing into recommended dimensions, such as $227 \times 227 \times 3$ or $224 \times 224 \times 3$ [4,5]. Available literature confirms the availability of traditional DLAs trained and validated with benchmark images [6–10]. Traditional DLAs, such as AlexNet, Visual Geometry Group-16 (VGG-16), VGG-19, Residual Network (ResNet) with various layer sizes such as 18, 34, 50, 101 and 152, and other traditional or customary architectures are chiefly considered to examine medical-grade images to support the automated disease detection process. The choice of a particular DLA depends on the speed of the diagnosis, required detection accuracy, and the availability of the computing facility. In this section, well known DLA such as AlexNet, VGG-16 and VGG-19 are implemented for the experimental demonstration. The disease detection procedures are executed using MATLAB and Python.

6.1 INTRODUCTION

The various stages of disease detection using medical images are presented in Figure 6.1. This figure also illustrates various techniques used in the literature to detect the diseases in medical images using a dedicated scheme.

The low-grade learning system is an early procedure adopted to detect the disease from a group of chosen images. The accuracy of this system was poor. This method detected diseases in a two-step process: feature extraction, and classification. The limitation of this technique is that, due to the two-step process, attaining accurate disease detection is less probable compared to the more recent ML technique.

147

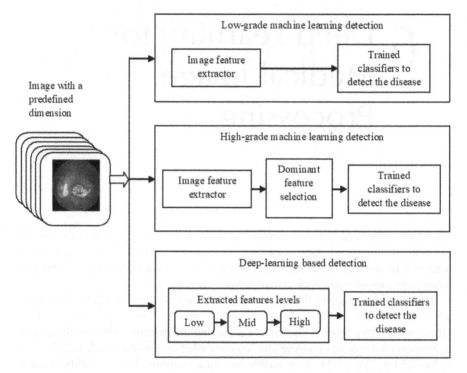

FIGURE 6.1   Existing Automated Disease Classification Techniques Available for Medical Images.

ML systems, also known as high-grade ML systems, help get better results due to improved schemes. This system includes a feature selection process, which helps attain higher detection accuracy while reducing the problem of over fitting. Because of this, the ML system is usually implemented by researchers to classify medical images. An added advantage of this system is that it can be implemented with smaller computation devices without sacrificing accuracy [11–13].

Although the DLA requires higher capacity computing facilities, it produces higher accuracy disease detection and works well on a range of greyscale and RGB-scale pictures irrespective of the imaging modality used. The main advantage of the DLA is, instead a customary network, transfer learning concept can be utilized to make use of the existing DL, as construction of a customary DLA system from the scratch is a time-consuming process. The training and testing procedures required in these systems are very high. The existing DLA based on transfer learning works well on disease detection problems and, in every case, it helps to attain better results. This DLA extracts the low-, mid-, and high-level features from each image and, based on the extracted information, will learn and remember. Further, all the extracted features are sorted based on their rank and a few features are discarded before it reaches the classifier (SoftMax). Classification accuracy attained with the deep classifier can be improved by implementing traditional classifiers available in the ML systems [14,15].

6.2 IMPLEMENTATION OF CNN FOR IMAGE ASSESSMENT

A breach in construction of networks for image categorization came with the finding that a Convolutional Neural Network (CNN) can be used to extract higher-level features from the test image. As a substitute to pre-processing the information to obtain features such as textures and shapes – as in an ML scheme – a CNN directly extracts the image's every unprocessed pixel information as input, and studies the information available in the test image to be processed. Normally, the CNN collects an input attribute map, a three-dimensional matrix in which the dimension of the first two magnitudes are related to the length and width of the image in pixels, and dimensions of the third is related to three-channel color components such as red, green, and blue – as in the case of RGB images – and gray pixel information – as in the case of greyscale images. The CNN encompasses a multitude of modules, each of which executes three functions. The working theory behind the CNN is explained below:

- **Convolution**

This procedure extracts features and generates an output feature map with different dimensions compared to the input feature map. This can be defined with two parameters:

i.   Dimension of the strips to be extracted (naturally 3×3 or 5×5 pixel map).
ii. The depth of the output feature map, which matches the amount of filters considered.

Throughout the convolution, the filters successfully glide over the input feature map's grid horizontally and vertically, one pixel at a time, extracting corresponding information from the test image.

The CNN executes element-wise reproduction of the filter and tile matrices for every filter-tile brace. It then sums all the rudiments of the resulting matrix to obtain a solitary charge. Each of these resulting values for each filter-tile pair is productively convoluted into a feature matrix.

Throughout preparation, the CNN discovers the finest values for the filter matrices to facilitate its mining of significant descriptions and information such as textures, edges, and shapes of the image using the values on the input feature map. The number of features extracted by the CNN rises when the number of filters used on the input increases. This increase in the filter size amplifies the training size, making the initial tuning of the CNN architecture more complex.

- **ReLU**

  Rectified Linear Unit (ReLU)-based alteration of the convolved feature is employed in CNN to introduce non-linearity into the model. ReLU function F $(r) = \max(0, r)$, proceeds r for all values of $r > 0$, and assigns 0 for all values of $r \leq 0$.
- **Pooling**

Pooling is executed in the CNN following ReLU. In this process, the CNN downsamples the convolved feature, which further drops the amount of aspects on the feature map. This process preserves the most significant feature data and is technically called the Max Pooling (MP) process. The MP works similar in style to convolution. The search begins with a tile with predefined shape which glides over the feature map to mine other tiles with a similar size. The maximum assessment is outputted to a new feature map for every tile, discarding all other values.

The operation of the MP is discussed below:

i. Size of the max-pooling filter (normally 2 × 2 pixels)
ii. Stride of the coldness in pixels after extrication of every tile

- **Fully Connected Layers**

The last part of a CNN includes one or more Fully Connected Layers (FCL) based on the need, in which every node is connected with other individual nodes of the network. The job of the FCL is to provide the essential one-dimensional feature vector to train, test, and validate the classifier unit found at the final part of the network. This classifier detects/classifies the medical images based on the selected features. To avoid the problem of over fitting, a discard operation is implemented to limit the image feature value. In most of the CNN, the classifier adopted is a SoftMax unit that performs a two-class/binary classification operation.

Figure 6.2 depicts the typical CNN with a single layer and, to achieve superior results, a considerable number of layers needs to be placed between the input image and the classifier system (SoftMax).

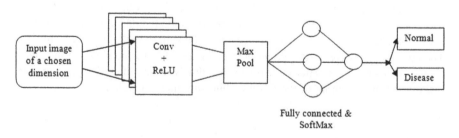

FIGURE 6.2 Simplified Structure of the CNN with Essential Operating Stages.

6.3 TRANSFER LEARNING CONCEPTS

Development of a CNN structure for a chosen task requires higher computational power (a computer with better RAM and graphics card) and additional time. The initial architecture construction needs more effort, and one cannot assure that a developed model is unique and will work on all image cases. To reduce the computational burden, the existing CNN (DLA) is adopted on medical images for the purpose of disease detection/classification.

Transfer learning is a machine-learning practice in which an existing and the pre-trained model of a DLA is utilized to perform disease detection operations. This section presents the results attained with simple pre-trained models (using the transfer learning approach), such as AlexNet, VGG-16 and VGG-19.

6.3.1 ALEXNET

AlexNet is one of the most successful and widely used CNN structures from 2012 onwards. It was proposed by Alex Krizhevsky [16]. AlexNet made a huge impact on the field of machine learning, particularly in the field of DL-to-machine vision. Essential information regarding the AlexNet can be found in [5–7]. This structure has eight layers: the first five include convolutional layers, followed by MP layers, and the last three indicates FCL as depicted in Figure 6.3. CNN uses the non-saturating ReLU activation utility, which demonstrates enhanced training performance over tanh and sigmoid utilities. Other essential information such as layer number, name, description, and dimension are clearly presented in Table 6.1. Due to its simple structure and higher classification accuracy, the AlexNet was widely considered for image classification tasks. This CNN could be executed using the MATLAB or Python softwares.

Along with the traditional AlexNet structures, a modified structure is proposed by the researchers. One improvement to the AlexNet with the deep- and handcrafted-feature is depicted in Figure 6.4. This structure will combine the learned features with the ML features to improve classification accuracy. Other information regarding this structure can be found in [5]. This structure will help to better the detection/classification of diseases by combining all possible image features.

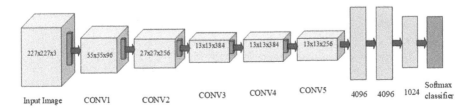

FIGURE 6.3 Pre-Trained AlexNet Architecture.

TABLE 6.1

Details of the CNN Architecture Used in AlexNet

Layer No.	Layer Name	Description	Dimension
1	Image	Accepts image input	227 × 227 × 3
2	Convolution	1-Convolution Layer	11 × 11 × 3 stride [4 4] padding [0 0 0 0]
3	ReLU	1-Rectified linear unit for removing negative values	–
4	Normalization	1-Normalization unit	–
5	Max-Pooling	1-Determines the Max value in a particular image array	3 × 3 with stride [2 2] and padding [0 0 0 0]
6	Convolution	2-Convolution layer	5 × 5 × 48 with stride [1 1] and padding [2 2 2 2]
7	ReLU	2-Rectified linear unit for removing negative values	–
8	Normalization	2-Normalization unit	–
9	Max-Pooling	2-Determines the Max value in a particular image array	3 × 3 max pooling with stride [2 2] and padding [0 0 0 0]
10	Convolution	3-Convolution layer	384 3 × 3 × 256 with stride [1 1] and padding [1 1 1 1]
11	ReLU	3-Rectified linear unit for removing negative values	–
12	Convolution	4-Convolution layer	384 3 × 3 × 192 convolutions with stride [1 1] and padding [1 1 1 1]
13	ReLU	4-Rectified linear unit for removing negative values	–
14	Convolution	5-Convolution layer	256 3 × 3 × 192 convolutions with stride [1 1] and padding [1 1 1 1]
15	ReLU	5-Rectified linear unit for removing negative values	–
16	Max Pooling	3-Determines the Max value in a particular image array	3 × 3 max pooling with stride [2 2] and padding [0 0 0 0]
17	Fully Connected	1-Fully connected artificial neural network	4096 fully connected layer
18	ReLU	6-Rectified linear unit for removing negative values	–
19	Dropout	1-Reduces the number of weights (50% dropout)	–
20	Fully Connected	2-Fully connected artificial neural network	4096 × 4096

(Continued)

TABLE 6.1 (Continued)

Layer No.	Layer Name	Description	Dimension
21	ReLU	7-Rectified linear unit for removing negative values	–
22	Dropout	2-Reduces the number of weights (50% dropout)	–
23	Fully Connected	3-Fully connected artificial neural network	4096 × 2
24	SoftMax	Activation layer	–
25	Classification Output	Normal/abnormal	–

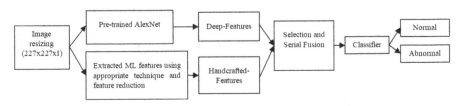

FIGURE 6.4   Enhanced AlexNet Architecture with Deep and Handcrafted Features.

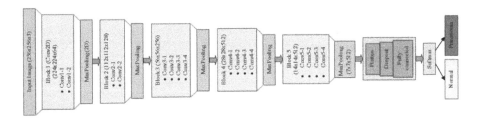

FIGURE 6.5   Pre-Trained VGG-16 Architecture.

6.3.2 VGG-16

VGG is a CNN architecture proposed by Simonyan and Zisserman [17] of the Visual Geometry Group (VGG) from the University of Oxford. The initial version VGG-16 features simple and effective predefined architecture and offers better results during the image classification task. The structure of the VGG-16 is depicted in Figure 6.5. The hardware description of the VGG-16 and its execution for brain image classification is presented in Table 6.2 and Figure 6.6, respectively.

TABLE 6.2

Details of the CNN Architecture Used in VGG-16

Layer (Type)	Output Shape	Parameter
input_1 (InputLayer)	(None, None, None, 3)	0
block1_conv1 (Conv2D)	(None, None, None, 64)	1792
block1_conv2 (Conv2D)	(None, None, None, 64)	36928
block1_pool (MaxPooling2D)	(None, None, None, 64)	0
block2_conv1 (Conv2D)	(None, None, None, 128)	73856
block2_conv2 (Conv2D)	(None, None, None, 128)	147584
block2_pool (MaxPooling2D)	(None, None, None, 128)	0
block3_conv1 (Conv2D)	(None, None, None, 256)	295168
block3_conv2 (Conv2D)	(None, None, None, 256)	590080
block3_conv3 (Conv2D)	(None, None, None, 256)	590080
block3_pool (MaxPooling2D)	(None, None, None, 256)	0
block4_conv1 (Conv2D)	(None, None, None, 512)	1180160
block4_conv2 (Conv2D)	(None, None, None, 512)	2359808
block4_conv3 (Conv2D)	(None, None, None, 512)	2359808
block4_pool (MaxPooling2D)	(None, None, None, 512)	0
block5_conv1 (Conv2D)	(None, None, None, 512)	2359808
block5_conv2 (Conv2D)	(None, None, None, 512)	2359808
block5_conv3 (Conv2D)	(None, None, None, 512)	2359808
block5_pool (MaxPooling2D)	(None, None, None, 512)	0
global_average_pooling2d_1	(None, 512)	0
dense_1 (Dense)	(None, 1024)	525312
dense_2 (Dense)	(None, 2)	2050

Total params: 15,242,050

Trainable params: 527,362

Non-trainable params: 14,714,688

```
Found 624 images belonging to 2 classes.
Found 156 images belonging to 2 classes.
Epoch 1/10
624/624 [==============================] - 61s 97ms/step - loss: 0.6279 - acc: 0.6667 - val_loss: 0.5584 - val_acc: 0.7436
Epoch 2/10
624/624 [==============================] - 23s 37ms/step - loss: 0.5582 - acc: 0.7268 - val_loss: 0.6292 - val_acc: 0.7051
Epoch 3/10
624/624 [==============================] - 23s 37ms/step - loss: 0.5441 - acc: 0.7416 - val_loss: 0.5420 - val_acc: 0.7564
Epoch 4/10
624/624 [==============================] - 23s 37ms/step - loss: 0.5404 - acc: 0.7336 - val_loss: 0.5209 - val_acc: 0.7372
Epoch 5/10
624/624 [==============================] - 23s 37ms/step - loss: 0.5234 - acc: 0.7388 - val_loss: 0.6049 - val_acc: 0.7308
Epoch 6/10
624/624 [==============================] - 23s 37ms/step - loss: 0.5247 - acc: 0.7444 - val_loss: 0.5234 - val_acc: 0.7436
Epoch 7/10
624/624 [==============================] - 23s 37ms/step - loss: 0.5177 - acc: 0.7376 - val_loss: 0.5072 - val_acc: 0.7436
Epoch 8/10
624/624 [==============================] - 23s 37ms/step - loss: 0.5196 - acc: 0.7424 - val_loss: 0.5116 - val_acc: 0.7500
Epoch 9/10
624/624 [==============================] - 23s 37ms/step - loss: 0.5160 - acc: 0.7472 - val_loss: 0.5047 - val_acc: 0.7436
Epoch 10/10
624/624 [==============================] - 23s 37ms/step - loss: 0.5093 - acc: 0.7452 - val_loss: 0.5040 - val_acc: 0.7692
```

FIGURE 6.6   Training Procedure Implemented for an Image Classification Task.

The input to cov1 layer is assigned to process an image with dimensions of 224 × 224 × 3 (RGB) and it also accepts images with dimensions of 224 × 224 × 1 (Gray). Image information is be extracted with filters of measuring 3 × 3. This value also utilizes the filter of dimension 1x1 at the end when the test image has passed through a mass of convolutional (Conv.) layers. The procedures, such as convolution and MP, continue as per the layers and, finally, a one-dimensional image feature vector reaches the FCL. This network has three layers of FCL and it is associated with the sorting and dropout functions. Through this process, a one-dimensional feature vector with dimensions of 1 × 1 × 1024 reaches the classifier section to train, test, and validate the function of SoftMax based on the considered image database. Earlier research works on the VGG-16 can be found in [4–7].

6.3.3 VGG-19

A modified form of the VGG-16 is the VGG-19. The construction and working principle is similar for both VGG-19 and VGG-16. The difference between them is illustrated in Figure 6.7. The function values used in VGG-19 are depicted in Table 6.3. The image-based training implemented with the VGG-19 architecture for a brain image classification task is presented in Figure 6.8.

As depicted in Figure 6.4, the performance of the VGG (VGG16/VGG19) architecture can be improved by integrating the deep and the machine learning features as illustrated in Figure 6.9. To integrate these features, serial or parallel feature fusion is employed as discussed in previous works [4,5].

6.4 MEDICAL IMAGE EXAMINATION WITH DEEP-LEARNING: CASE STUDY

To demonstrate the image examination capability of the DLA, an experimental investigation is implemented using the MATLAB/Python software. The results are also presented. During this study, the results attained with the SoftMax classifier alone and the superiority of the considered DLA is judged by constructing the Confusion-Matrix (CM). The typical CM considered in the literature is illustrated in Figure 6.10. Based on the values of the TP, TN, FP, and FN, essential performance measures, such as ACC, PRE, SEN, SPE and NPV, are computed. Superiority of this value will confirm the performance of the architecture.

6.4.1 Brain Abnormality Detection

Abnormalities in the brain severely affect the whole body, hence the severity and orientation must be detected through imaging techniques. MRI with a chosen modality is adopted by doctors for brain abnormality detection. The disease in the brain MRI slice can be assessed by a radiologist and a doctor. The evaluation of the brain condition using the MRI is a time consuming process thus computerised schemes are widely proposed and implemented. The DLA-based brain examination is proposed and implemented in research, and the proposed work presents the MRI-

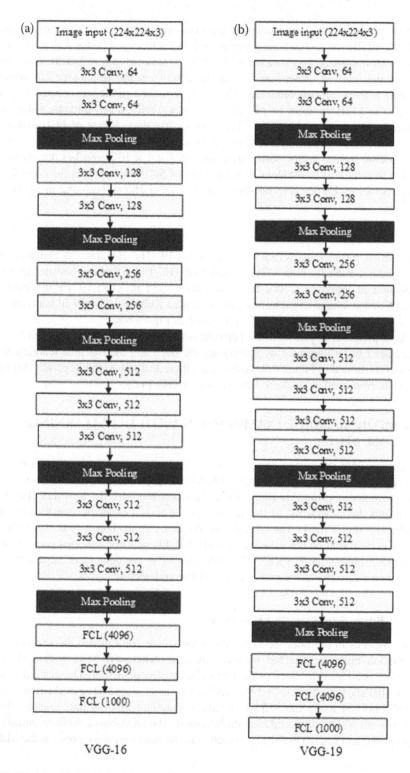

FIGURE 6.7 VGG Architectures.

TABLE 6.3

Details of the CNN Architecture Used in VGG-19

Layer (Type)	Output Shape	Parameter
input_1 (InputLayer)	(None, None, None, 3)	0
block1_conv1 (Conv2D)	(None, None, None, 64)	1792
block1_conv2 (Conv2D)	(None, None, None, 64)	36928
block1_pool (MaxPooling2D)	(None, None, None, 64)	0
block2_conv1 (Conv2D)	(None, None, None, 128)	73856
block2_conv2 (Conv2D)	(None, None, None, 128)	147584
block2_pool (MaxPooling2D)	(None, None, None, 128)	0
block3_conv1 (Conv2D)	(None, None, None, 256)	295168
block3_conv2 (Conv2D)	(None, None, None, 256)	590080
block3_conv3 (Conv2D)	(None, None, None, 256)	590080
block3_conv4 (Conv2D)	(None, None, None, 256)	590080
block3_pool (MaxPooling2D)	(None, None, None, 256)	0
block4_conv1 (Conv2D)	(None, None, None, 512)	1180160
block4_conv2 (Conv2D)	(None, None, None, 512)	2359808
block4_conv3 (Conv2D)	(None, None, None, 512)	2359808
block4_conv4 (Conv2D)	(None, None, None, 512)	2359808
block4_pool (MaxPooling2D)	(None, None, None, 512)	0
block5_conv1 (Conv2D)	(None, None, None, 512)	2359808
block5_conv2 (Conv2D)	(None, None, None, 512)	2359808
block5_conv3 (Conv2D)	(None, None, None, 512)	2359808
block5_conv4 (Conv2D)	(None, None, None, 512)	2359808
block5_pool (MaxPooling2D)	(None, None, None, 512)	0
global_average_pooling2d_1	((None, 512)	0
dense_1 (Dense)	(None, 1024)	525312
dense_2 (Dense)	(None, 2)	2050

Total params: 20,551,746

Trainable params: 527,362

Non-trainable params: 20,024,384

based brain tumor detection implemented with a chosen image database. This work deploys binary classification to separate the brain tumor database into normal and abnormal class. A multi-class separation is also implemented to separate the database into low-grade (LGG) tumor, high-grade tumor (HGG), stroke lesion, and the normal class using a multi-class detection system. Information regarding the image database and the results with the MATLAB software, AlexNet, and VGG-16 DLA is presented in this section.

This work considered the transfer-learning technique. The datasets were resized to 227 × 227 × 1 pixels before the implementation. The number of images considered for the assessment is depicted in Table 6.4, and the sample test images

```
Found 624 images belonging to 2 classes.
Found 156 images belonging to 2 classes.
Epoch 1/10
624/624 [==============================] - 84s 134ms/step - loss: 0.6350 - acc: 0.6643 - val_loss: 0.5140 - val_acc: 0.7179
Epoch 2/10
624/624 [==============================] - 28s 45ms/step - loss: 0.5686 - acc: 0.7264 - val_loss: 0.5197 - val_acc: 0.7500
Epoch 3/10
624/624 [==============================] - 29s 46ms/step - loss: 0.5455 - acc: 0.7376 - val_loss: 0.5054 - val_acc: 0.7500
Epoch 4/10
624/624 [==============================] - 29s 46ms/step - loss: 0.5349 - acc: 0.7400 - val_loss: 0.5085 - val_acc: 0.7628
Epoch 5/10
624/624 [==============================] - 28s 46ms/step - loss: 0.5259 - acc: 0.7472 - val_loss: 0.6038 - val_acc: 0.7436
Epoch 6/10
624/624 [==============================] - 28s 46ms/step - loss: 0.5279 - acc: 0.7328 - val_loss: 0.5244 - val_acc: 0.7628
Epoch 7/10
624/624 [==============================] - 28s 46ms/step - loss: 0.5162 - acc: 0.7556 - val_loss: 0.5220 - val_acc: 0.7628
Epoch 8/10
624/624 [==============================] - 28s 46ms/step - loss: 0.5091 - acc: 0.7468 - val_loss: 0.5054 - val_acc: 0.7436
Epoch 9/10
624/624 [==============================] - 28s 46ms/step - loss: 0.5050 - acc: 0.7516 - val_loss: 0.5071 - val_acc: 0.7308
Epoch 10/10
624/624 [==============================] - 28s 46ms/step - loss: 0.5129 - acc: 0.7488 - val_loss: 0.5291 - val_acc: 0.7756
```

FIGURE 6.8   Training Procedure Implemented Using VGG-19 for an Image Classification Task.

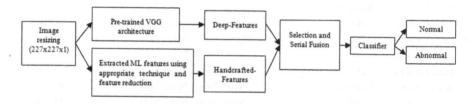

FIGURE 6.9   Improving the Classification Accuracy of the VGG Using Deep and Handcrafted Features.

	Target value		
	True Positive (TP)	False Positive (FP)	Sensitivity (SEN)
	False Negative (FN)	True Negative (TN)	Specificity (SPE)
	Precision (PRE)	Negative Predicted Value (NPV)	Accuracy (ACC)

(Attained value — left axis label)

FIGURE 6.10   Typical Confusion-Matrix Structure to Appraise the Performance.

considered in this work are shown in Figure 6.11. In this work, the size of the image dataset is increased with the help of the image augmentation technique, which increases the test image size with rotation, horizontal flip, and vertical flip.

TABLE 6.4

Brain MRI Slices Considered for the Assessment

Image Class	Modality	Dimension	Number of Images (Training)	Number of Images (Validation)
Low Grade Glioma (LGG)	T2	2D	400	100
High Grade Glioma (HGG)	T2	2D	400	100
Ischemic stroke	Flair	2D	400	100
Normal	Flair + T2	2D	400	100
	Total		1600	400

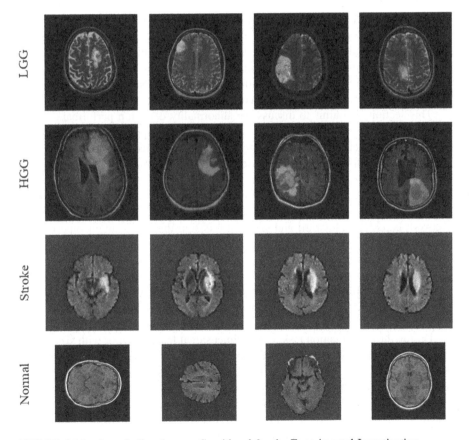

FIGURE 6.11 Sample Test Images Considered for the Experimental Investigation.

The detailed procedure on the AlexNet and its implementation for image classification can be found in [18,19].

The attained result for the two-class and the multi-class problems are depicted with the help of the CM and, based on the attained accuracy, performance is validated.

Initially, a binary classification is executed with the traditional AlexNet architecture and the attained results were presented. During this process, the image results obtained from all the convolution and max-pooling layers were extracted separately. These values are illustrated in Figures 6.12 and 6.13, respectively. Figure 6.14 presents the experimental outcome attained with the test images (normal and augmented) and the corresponding CM is depicted in Figure 6.15. It can be noted from the results that the proposed procedure helped attain a classification accuracy of 96.6%. This value can be improved by replacing the SoftMax classifier with other existing classifiers or by integrating the deep features with the handcrafted features as depicted in Figure 6.4.

After executing the binary classification task, the dataset then undergoes a multi-class classification. The result is depicted in Figures 6.16 and 6.17. Figure 6.16 shows the accuracy and the loss function obtained during the training, testing, and validation task and the corresponding CM is depicted in Figure 6.17. The AlexNet with transfer-learning concept helped attain a classification accuracy of 99.3% for a multi-class problem. This result confirms that the AlexNet works well to detect brain abnormalities from the MRI recorded with modalities such as Flair and T2.

This section also aims to discuss the abnormality detection performance of the pre-trained VGG-16/VGG-19 network, which aims to improve detection accuracy. This work considers the classification of the brain MRI slices into LGG/HGG using the VGG-19 architecture and the essential experimental investigation is verified using the MATLAB-based simulation.

Necessary information on the VGG-19 design and its applications can be found in [15,16]. Previous DLA-assisted image processing confirmed that the VGG-19 network is a straightforward and successful design that helps attain superior classification results. Further, this architecture requires less computation effort compared to the other architectures.

2D brain MRI slices with tumor grades LGG and HGG were initially resized into $224 \times 224 \times 1$ pixels, followed by an augmentation procedure to augment the number of imagery. Table 6.5 presents the number of training and validation images of the T2 modality under study.

During the classification task, a one-dimensional feature vector with size $1 \times 1 \times 1000$ is considered to train, test, and validate the SoftMax classifier associated with the VGG-19. A five-fold cross-validation is employed and the best value attained is considered as the final classification result.

The VGG-19 has various layers. The result attained for the MaxPool layers of the first two stages is depicted in Figure 6.18 and the corresponding search pattern and attained CM are depicted in Figures 6.19 and 6.20, respectively. In this work, the classification is verified with a ten-fold cross-validation. The SoftMax helped achieve a classification accuracy of >93%. This accuracy can be enhanced by

Fully Connected Layer (FC-1) Network Activation for 25 Hidden Neurons

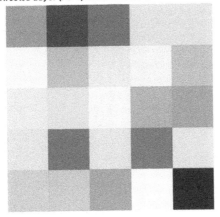

ConvNet Layer-2 for 4 Channels

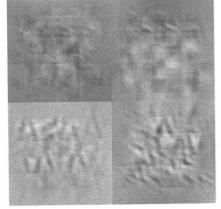

ConvNet Layer-3 for 4 Channels

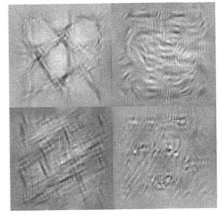

ConvNet Layer-4 for 4 Channels

ConvNet Layer-5 for 4 Channels

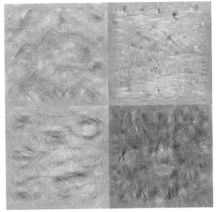

FIGURE 6.12 The Results Attained from Every Convolution Layer with Affixed Channels.

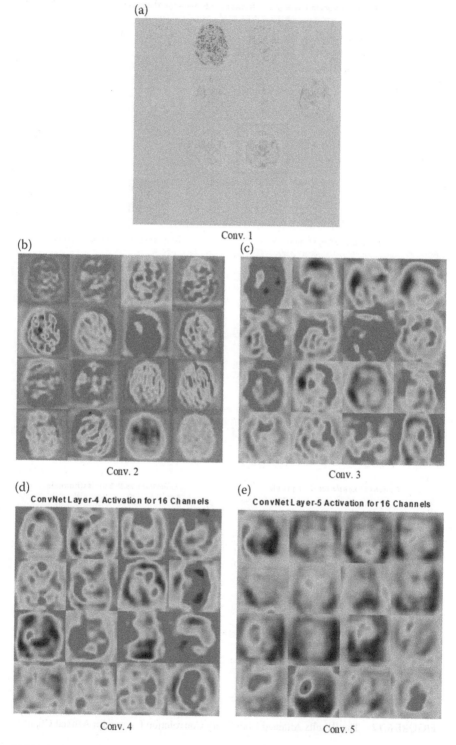

(a)

Conv. 1

(b)

Conv. 2

(c)

Conv. 3

(d)

ConvNet Layer-4 Activation for 16 Channels

Conv. 4

(e)

ConvNet Layer-5 Activation for 16 Channels

Conv. 5

FIGURE 6.13   The Results Attained for a Chosen Test Image from Each Convolution Layer.

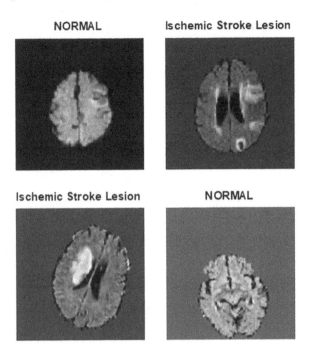

FIGURE 6.14 Two Class Classification Applied to Get Normal/Stroke MRI.

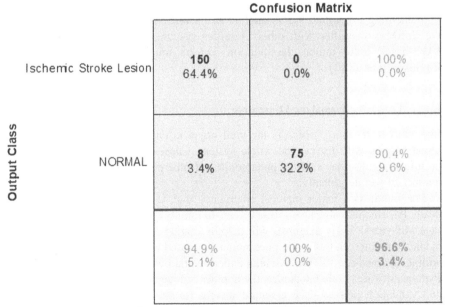

FIGURE 6.15 Confusion Matrix of Two Class Classification Operation.

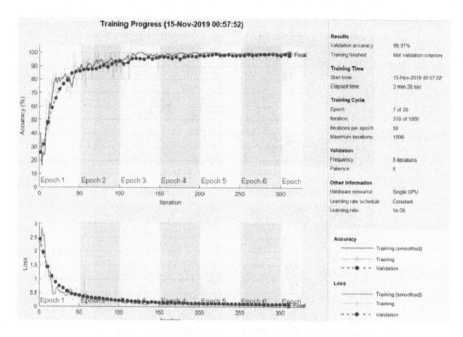

FIGURE 6.16 The Training, Testing and Validation Result Attained During the Multi-Class Classification.

implementing a modified VGG structure as discussed in Figure 6.9 or by replacing the SoftMax classifier with other classifier systems. Additional details on the VGG-based brain tumor classification can be found in the recent work of Rajinikanth et al. [4].

6.4.2 LUNG ABNORMALITY DETECTION

Infections in the lung, which is the vital organ in charge of supplying air to the blood stream, will distress the whole system. Causes of lung abnormality could be infectious diseases such as pneumonia and tuberculosis, which affects a large number of people globally.

Pneumonia is a major medical emergency that, if left untreated, may lead to death. Pneumonia will cause severe effects in infants (age <5 years) and the elderly (age >65 years). Early diagnosis will help in treating those affected with improved medical facilities. In this work, pneumonia diagnosed with the AlexNet is presented with its attained results. The essential experimental investigation is implemented in Python software. Table 6.6 depicts the number of images considered for the training and validation process. The essential images for this study were obtained from [20,21]. Sample test images considered in this work are shown in Figure 6.21. The sample results obtained from various layers of DLA are presented in Figures 6.22 and 6.23. They present the accuracy and loss function attained during the training and validation process. The sample classification results during the validation

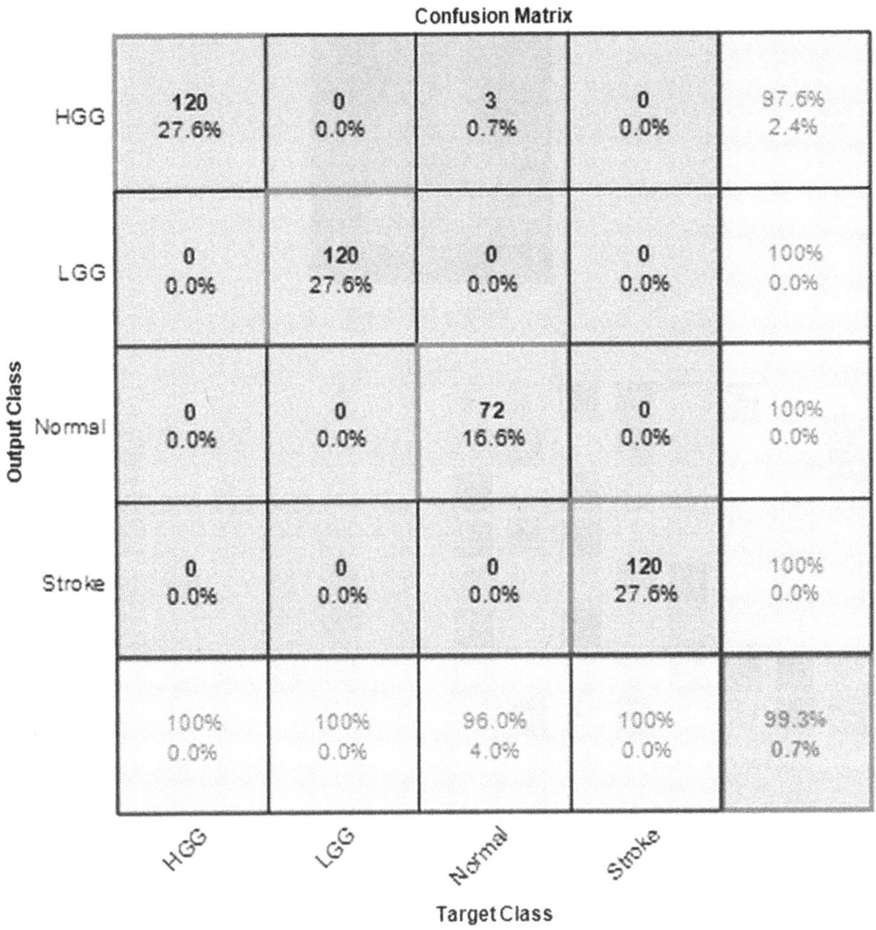

FIGURE 6.17   Confusion Matrix Obtained for Multi-Class Classification.

TABLE 6.5
Number of Images Considered to Test the Performance of VGG-16

Image Class	Pixel Dimension	Number of Images for Training	Number of Images for Validation
LGG	224 × 224 × 1	600	200
HGG	224 × 224 × 1	600	200

(a)

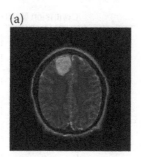

Sample test image

(b)   (c)

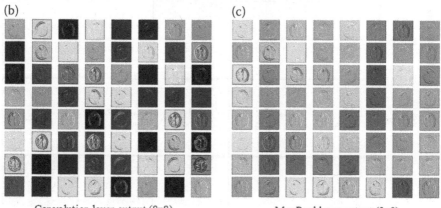

Convolution layer output (8x8)   MaxPool layer output (8x8)

FIGURE 6.18   The Results Attained from the First Layer of the VGG=19 DLA.

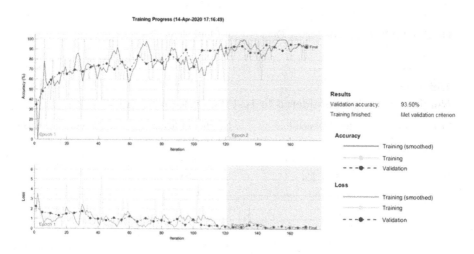

FIGURE 6.19   Convergence of the Training and Validation Operation.

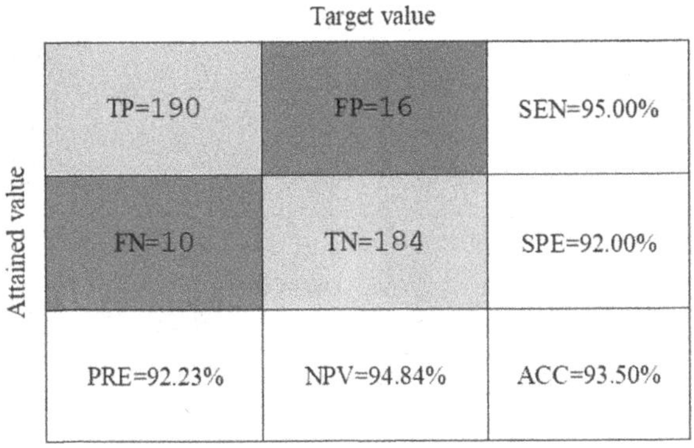

Target value

TP=190	FP=16	SEN=95.00%
FN=10	TN=184	SPE=92.00%
PRE=92.23%	NPV=94.84%	ACC=93.50%

Attained value

FIGURE 6.20   Confusion Matrix Attained with the VGG-19 DLA.

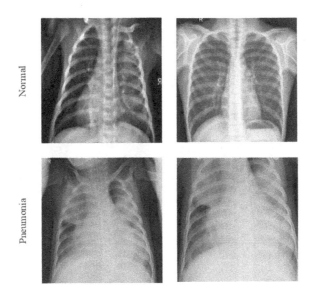

Normal

Pneumonia

FIGURE 6.21   Sample Test Images Considered in This Study.

TABLE 6.6

Total Number of Test Images of Chest X-Ray Considered for the Disease Detection Process

Image Class	Image Dimension	Number of Images for Training	Number of Images for Testing
Normal	227 × 227 × 1	600	200
Pneumonia	227 × 227 × 1	600	200
	Total	1200	400

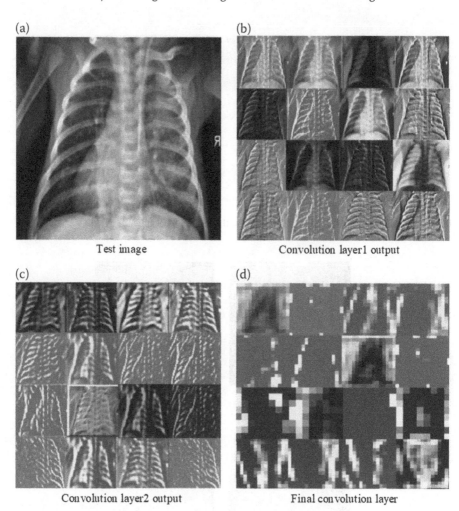

FIGURE 6.22   Results Obtained for a Chosen Test Image.

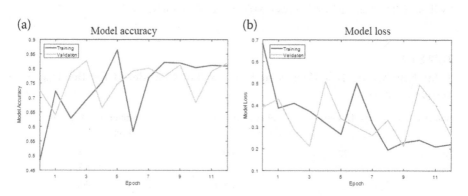

FIGURE 6.23   The Results Attained with the Training and Testing Process with VGG-19.

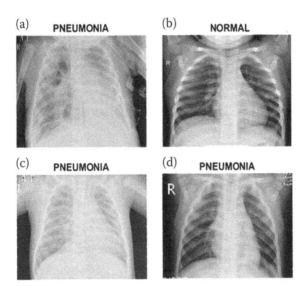

FIGURE 6.24   Validation Results Attained for the Pneumonia Test Images.

Target value

TP=189	FP=8	SEN=94.50%
FN=11	TN=192	SPE=96.00%
PRE=95.94%	NPV=94.84%	ACC=94.58%

Attained value

FIGURE 6.25   Confusion Matrix Attained for the Pneumonia Classification Problem.

process and the confusion matrix are presented in Figures 6.24 and 6.25, respectively. The result of this study shows that the proposed DLA helps achieve a classification accuracy of 94.58%.

A similar database is then examined using the VGG-16 architecture with both MATLAB and Python software. The results are depicted in the discussion. Figure 6.26 presents the results from the various layers of the DLA and Figure 6.27 illustrates the activation function. Figure 6.28 depicts the confusion matrix, which shows that the classification accuracy attained with the SoftMax classifier is 96.30%.

The results from the AlexNet architecture with Python software-based simulation are presented for the dataset of Table 9. Figure 6.29 presents the convergence of the accuracy and the loss functions while Figure 6.30 shows the outcome attained

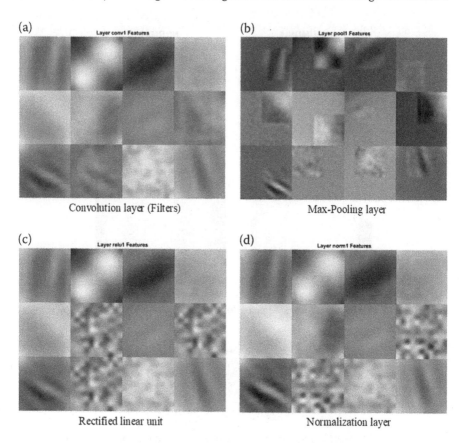

FIGURE 6.26   Results Attained from the Layers.

FIGURE 6.27   Activation Maps of the Convolution Layer (conv1) for a Sample X-Ray Image.

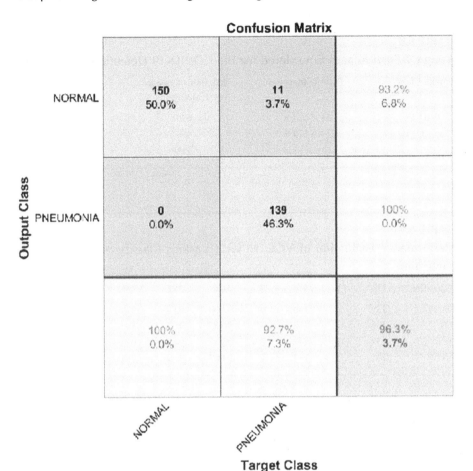

FIGURE 6.28 Confusion Matrix Attained with the AlexNet Architecture.

TABLE 6.7
Fundus Test Images Considered in This Study

Image Class	Pixel Dimension	Test Images	Test Images after Augmentation	Number of Images for Validation
AMD	227 × 227 × 3	400	400 × 3 = 1200	150
Normal	227 × 227 × 3	400	400 × 3 = 1200	150
	Total	800	2400	300

TABLE 6.8

Lung CT Scan Images Considered for the COVID-19 Detection Process

Image Class	Pixel Dimension	Number of Images for Training	Number of Images for Validation
COVID-19	224 × 224 × 1	600	150
Non-COVID-19	224 × 224 × 1	600	150
Total		1200	300

TABLE 6.9

Performance Evaluation of VGG-16 with Various Classifiers for COVID-19 Detection

Classifier	TP	FN	TN	FP	ACC	PRE	SEN	SPE	NPV
SoftMax	139	11	140	10	93.00	93.28	92.67	93.33	92.71
DT	133	17	138	12	90.33	91.72	88.67	92.00	89.03
KNN	139	11	142	8	93.67	94.56	92.67	94.67	92.81
SVM	141	9	143	7	94.67	95.27	94.00	95.33	94.08

(a)   (b)

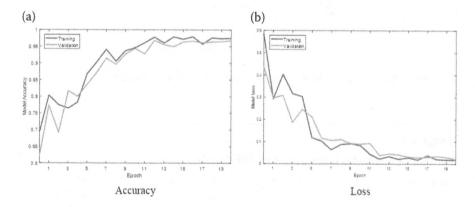

Accuracy   Loss

FIGURE 6.29 Accuracy and Loss Function Attained with the AlexNet Architecture.

from various convolution layers. The confusion matrix shown in Figure 6.31 confirms that proposed work helped achieve a classification accuracy of 95.5%. These results confirm that the DLA, such as VGG and the AlexNet, works well on medical image classification irrespective of the software used, such as the

FIGURE 6.30   The Results Attained in the Various Convolution Layers.

Target value

TP= 194	FP=12	SEN=97.00%
FN=6	TN= 188	SPE=94.00%
PRE= 94.17%	NPV= 96.91%	ACC= 95.50%

Attained value

FIGURE 6.31   Confusion Matrix Attained with AlexNet during the Pneumonia Classification.

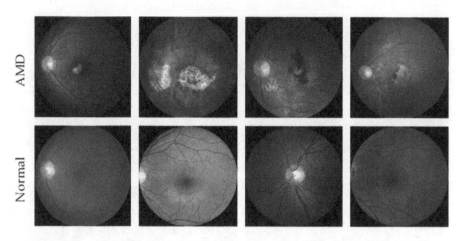

FIGURE 6.32   Sample Retinal Test Images of the AMD/Normal Class.

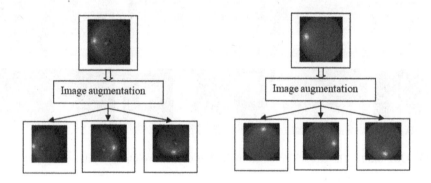

FIGURE 6.33   Increasing the Database Amount with Data Augmentation.

MATLAB and Python. The choice of a particular DLA and the software mainly depends on the user's expertise.

6.4.3 RETINAL ABNORMALITY DETECTION

The eye is another vital organ, diseases of which require special care. Diseases in the retina can be examined using the images recorded with a fundus camera. This section presents the AlexNet-based classification of the retinal abnormality known as Age-related Macular Degeneration (AMD). The number of test images considered in this work is depicted in Table 6.7 and the sample test images considered in this work are shown in Figure 6.32. The number of images available in the dataset is lower, thus data augmentation is implemented to improve the image size through vertical and horizontal rotation. The process is depicted in Figure 6.33.

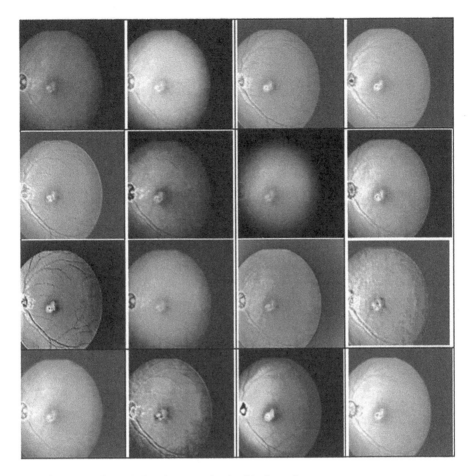

FIGURE 6.34 Convolution Outcome Attained in Layer1.

Figures 6.34 and 6.35 illustrate the outcome of the convolution layers of the AlexNet, and similar results are attained for other layers of the AlexNet. The classification result attained with the SoftMax classifier is depicted in Figure 6.36, confirming that the proposed DLA helped attain an accuracy of 63.67% for the considered retinal image database.

6.4.4 COVID-19 Lesion Detection

Recently, the respiratory tract infection called COVID-19 infected the global population, which has already caused a number of deaths within a short span of time. COVID-19 is due to the Severe Acute Respiratory Syndrome-Corona Virus-2 (SARS-CoV-2), which causes acute pneumonia and moderate to severe respiratory problems. The untreated disease has caused death especially among the elderly.

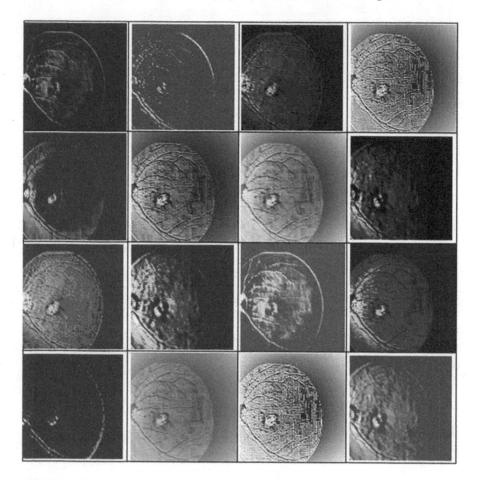

FIGURE 6.35 Convolution Outcome Attained in Layer2.

Target value

TP= 142	FP=11	SEN= 94.67%
FN=8	TN= 139	SPE= 92.67%
PRE= 92.81%	NPV= 94.56%	ACC= 93.67%

Attained value

FIGURE 6.36 Confusion Matrix Attained with the AlexNet for the Retinal Database.

Due to its rapidity and death rate, it was declared a pandemic by the World Health Organization (WHO). Scientists are working to find solutions to stop its impact on the community. Due to its significance, a number of disease detection procedures were proposed by researchers, such as the image-assisted COVID-19 detection technique. The lung section recorded with the chest X-ray or the CT scan are assessed to identify the presence and severity of COVID-19 in a patient. The doctors can then plan appropriate treatment procedures to reduce the possibility of death when the diagnosis is accurate. The lung CT scan-assisted detection is a procedure and, in this section, the assessment of COVID-19 from the CT scan slices is presented in brief.

This section presents the initial procedures to be followed in implementing a chosen DLA using the MATLAB.

Step 1: Collect the clinical-grade lung CT scan images (CTI) from hospitals or from the benchmark image datasets. The essential CTI for research purposes can be found in [22,23].

Step 2: Extract the 2D slices and resize to an appropriate value based on the DLA to be executed. The commonly accepted image dimension are $227 \times 227 \times 3$ or $224 \times 224 \times 3$.

Step 3: If are less images, then implement image augmentation and improve the test image size and store the images (normal/disease case) in a separate location.

Step 4: Chose an appropriate DLA and assign the initial parameters, such as the number of iterations, number of epochs, refreshment/updating rate and the error rate, etc.

Step 5: Implement the training, testing, and validation operations. If the results are appropriate, use the current result, else, repeat the procedure and allow the DLA to learn the new database to be classified.

Step 6: After sufficient learning, execute the procedure and validate the performance of the pre-trained modal with a new dataset. Record the result.

COVID-19 lesion detection using the CTI is implemented with the VGG-16 architecture. The test images considered in this study and its available locations are presented in Table 6.8 while the sample test images are depicted in Figure 6.37. There are less images ($500 + 500 = 1000$) hence, data augmentation (image flip) is implemented to increase the total size of the images to 2000 (1000 original + 1000 flipped). These images are then considered to train, test, and validate the VGG-16 architecture (Transfer Learning technique) implemented using the MATLAB software.

Before initiating the DLA training process for the considered CTI database, the essential values such as the maximum epochs, iteration, error rate, and updating are assigned. The VGG-16 architecture is allowed to learn from every image based on the attained features. During the learning process, every layer in the VGG-16 passes the processed image to the next layer with the help of Max-pooling. This process continues until the essential features of an image ($1 \times 1 \times 1000$) are obtained for the classifier. The results of the layers ($8 \times 8 = 64$) attained during the training process is

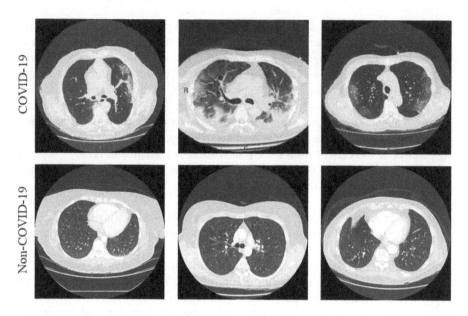

FIGURE 6.37   Sample Test Images Considered in This Study.

(a)

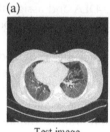

Test image

(b)   (c)

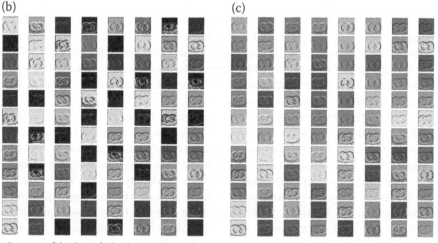

Outcome of the Convolution layer 1 (8x8=64)   Outcome of the Maxpool layer 1 (8x8=64)

FIGURE 6.38   Sample Result Obtained with the DLA during the Training Process.

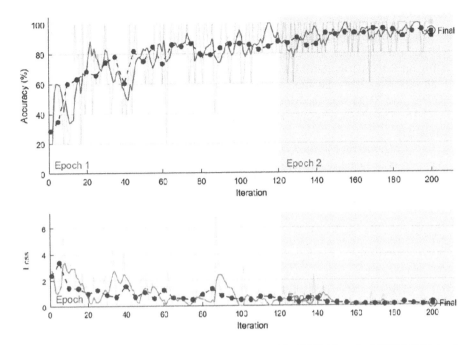

FIGURE 6.39   Training and Validation Performance of the VGG-16 on the COVID-19 Dataset

Target value

TP= 139	FP=10	SEN= 92.67%
FN=11	TN= 140	SPE= 93.33%
PRE= 93.28%	NPV= 92.71%	ACC= 93.00%

Attained value

FIGURE 6.40   Confusion Matrix Attained with VGG-16 with SoftMax Classifier for COVID-19 Detection.

presented in Figure 6.38 (for the first layer) and similar results are obtained for other layers with dimensions of 128, 256, 512, and 512, respectively. This procedure is repeated for every single image in the dataset. Finally the DLA learns about the image based on the existing features and the learning rate obtained is presented in Figure 6.39. After training the DLA with chosen test images, its detection performance is then validated. Based on the attained accuracy, the disease detection performance of VGG-16 is confirmed. If the first trial produces a result with lower accuracy, then the training procedure is to be repeated until the DLA learns completely and offers better classification accuracy.

The commonly used classifier in the DLA is the SoftMax but, in some cases, the classification performance can be improved by replacing the SoftMax with other well-known classifiers. The classification accuracy attained in this work is depicted in Figure 6.40 and the SoftMax helped attain an accuracy of 93%.

In this work, the classifiers such as Random Forest (RF), k-nearest neighbor (KNN), and SVM were considered to enhance the classification accuracy of the VGG-16. The attained results were then compared to the SoftMax classifier for validation.

Initial parameters assigned for the DLA before beginning the disease detection operation is depicted in the following table:

Parameter	Value
Number of epochs	10
Number of iterations	1000
Iterations per epoch	100
Updating frequency during validation	5 iterations
Learning rate schedule	Constant
Error value	0.0001
Hardware description	Core i5 processor with 8 GB RAM and 4GB graphics card
Software description	MATLAB

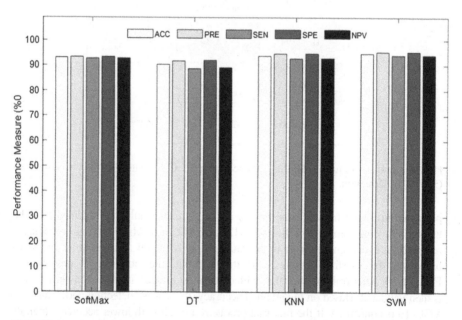

FIGURE 6.41   Performance Evaluation of VGG-16 for the COVID-19 Database.

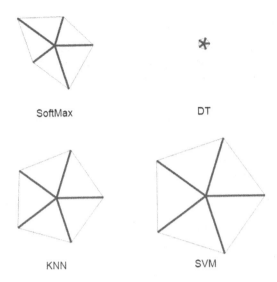

SoftMax   DT

KNN   SVM

FIGURE 6.42   Glyph Plot to Evaluate the Overall Performance of Employed Classifiers.

The classification results attained with DT, KNN, and the SVM with linear kernel are presented in Table 6.9. Its graphical representation is depicted in Figure 6.41. This result confirms that the VGG-16 with the SVM classifier obtain better abnormality detection accuracy compared to other classifiers. Further, the accuracy of DT is poor compared to the SoftMax. To assess the overall performance of the classifier, a glyph-plot (Figure 6.42) is constructed. This plot confirms that the VGG-16 together with the SVM classifier helps attain better overall performance compared to other classifiers considered in this study. The maximum classification accuracy attained with this study is 94.67%. This value can be further improved by implementing a modified VGG-16 architecture by combining the deep-features with the handcrafted-features as discussed in [4,5].

6.5 SUMMARY

This section presents an overview of the CNN-based abnormality classification system using medical images. This section also discussed an overview of Deep-Learning systems such as AlexNet, VGG-16, and VGG-19. The concept of the transfer-learning technique is presented with appropriate experimental results and the validation of the obtained results based on the confusion matrix are also discussed. Further, this section presented a guiding procedure to improve the performance of the deep-learning system with deep and handcrafted features. The need for training, testing, and validation of the binary and multi-class classifiers are also discussed with results attained with MATLAB and Python software.

REFERENCES

1. Hoo-Chang, S., Roth, H.R., Gao, M., Lu, L., Xu, Z., Nogues, I., Yao, J., Mollura, D. & Summers, R.M. (2016). Deep convolutional neural networks for computer-aided detection: CNN architectures, dataset characteristics and transfer learning. *Transactions on Medical Imaging*, 35, 1285–1298.
2. Zhou, M., Tian, C., Cao, R., Wang, B., Niu, Y., Hu, T., Guo, H. & Xiang, J. (2018). Epileptic seizure detection based on EEG signals and CNN. *Frontiers in Neuroinformatics*, 12, 95.
3. Yousefi, M., Krzyżak, A. & Suen, C.Y. (2018). Mass detection in digital breast tomosynthesis data using convolutional neural networks and multiple instance learning. *Computers in Biology and Medicine*, 96, 283–293.
4. Rajinikanth, V., Raj, A.N.J., Thanaraj, K.P. & Naik, G.R. (2020). A customized VGG19 network with concatenation of deep and handcrafted features for brain tumor detection. *Applied Sciences*, 10(10), 3429.
5. Bhandary, A., et al. (2019). Deep-learning framework to detect lung abnormality–A study with chest X-ray and lung CT scan images. *Pattern Recognition Letters*, 129, 271–278.
6. Arshad, H., Khan, M.A., Sharif, M.I., Yasmin, M., Tavares, J.M.R.S., Zhang Y.-D. & Satapathy, S.C. (2020). A multilevel paradigm for deep convolutional neural network features selection with an application to human gait recognition. *Expert Systems*, e12541.
7. Khan, M.A., Akram, T., Sharif, M., Awais, M., Javed, K., Ali, H., et al. (2018). CCDF: automatic system for segmentation and recognition of fruit crops diseases based on correlation coefficient and deep CNN features. Computers and Electronics in Agriculture, 155, 220–236.
8. Khan, M.A., Khan, M.A., Ahmed, F., Mittal, M., Goyal, L.M., Hemanth, D.J., et al. (2020). Gastrointestinal diseases segmentation and classification based on duo-deep architectures. *Pattern Recognition Letters*, 131, 193–204.
9. Rashid, M., Khan, M.A., Sharif, M., Raza, M., Sarfraz, M.M., Afza, F.J.M.T., et al. (2019). Object detection and classification: A joint selection and fusion strategy of deep convolutional neural network and SIFT point features, 78, 15751–15777.
10. Rehman, A., Khan, M.A., Mehmood, Z., Saba, T., Sardaraz, M. & Rashid, M. (2020). Microscopic melanoma detection and classification: A framework of pixel-based fusion and multilevel features reduction. *Microscopy Research and Technique*, 83(4), 410–423.
11. Zuo, H., Fan, H., Blasch, E. & Ling, H. (2017). Combining convolutional and recurrent neural networks for human skin detection. *IEEE Signal Processing Letters*, 24, 289–293.
12. Charron, O., Lallement, A., Jarnet, D., Noblet, V., Clavier, J.-B. & Meyer, P. (2018). Automatic detection and segmentation of brain metastases on multimodal MR images with a deep convolutional neural network. *Computers in Biology and Medicine*, 95, 43–54.
13. Khan, M.A., Akram, T., Sharif, M., Javed, M.Y., Muhammad, N. & Yasmin, M. (2019). An implementation of optimized framework for action classification using multilayers neural network on selected fused features. *Pattern Analysis and Applications*, 22, 1377–1397.
14. Cao, X., Wu, C., Yan, P. & Li, X. (2011). Linear SVM classification using boosting HOG features for vehicle detection in low-altitude airborne videos. In: *2011 18th IEEE International Conference on Image Processing*, 2421–2424.
15. Sharif, M., Khan, M.A., Tahir, M.Z., Yasmim, M., Saba, T. & Tanik, U.J. (2020). A Machine Learning Method with Threshold Based Parallel Feature Fusion and Feature

Selection for Automated Gait Recognition. *Journal of Organizational and End User Computing (JOEUC)*, 32(2), 67–92.

16. Krizhevsky, A., Sutskever, I. & Hinton, G.E. (2012). ImageNet classification with deep convolutional neural networks. *Advances in Neural Information Processing Systems*, 25, 1097–1105.

17. Simonyan, K. & Zisserman, A. (2014). Very deep convolutional networks for large-scale image recognition. arXiv:1409.1556 [cs.CV].

18. Talo, M., Baloglu, U.B., Yıldırım, Ö. & Acharya, U.R. (2019). Application of deep transfer learning for automated brain abnormality classification using MR images. Cognitive Systems Research, 54, 176–188.

19. Baloglu, U.B., Talo, M., Yildirim, O., Tan, R.S. & Acharya, U.R. (2019). Classification of myocardial infarction with multi-lead ECG signals and deep CNN. *Pattern Recognition Letters*, 122, 23–30.

20. Rajpurkar, P., et al. (2017). CheXNet: radiologist-level pneumonia detection on chest x-rays with deep learning. arXiv:1711.05225 [cs.CV].

21. Irvin, J., et al. (2019). CheXpert: a large chest radiograph dataset with uncertainty labels and expert comparison. arXiv:1901.07031 [cs.CV].

22. http://medicalsegmentation.com/covid19/ (Accessed on: 18 May 2020).

23. Jun, M., Cheng, G., Yixin, W., Xingle, A., Jiantao, G., et al. (2020). COVID-19 CT Lung and Infection Segmentation Dataset (Version Verson 1.0) [Data set]. *Zenodo*. doi: 10.5281/zenodo.3757476.

7 Conclusion

Due to a variety of reasons, health abnormalities are gradually rising while the improved medical and computing facilities work to control diseases with various techniques such as (i) premature detection, (ii) automated detection of the disease with invasive and non-invasive samples, (iii) assisting the doctors during the treatment planning and execution and (iv) monitoring the patients during the pre- and post-treatment process.

Diseases in internal organs are more acute compared to those of the external organs. Identification of these illnesses with better accuracy needs advanced diagnostic techniques, which helps to identify the disease with through tissue samples, blood samples, bio-signals, and bio-images. After collecting the essential information from the patients, it is the responsibility of the doctor to assess these samples with care and plan for essential treatment to control/cure the disease. Bio-imaging is one of the non-invasive techniques commonly preferred to record the images of the internal organs to examine the disease and its severity with considerable accuracy. Recent developments in science helped implement a range of imaging modalities to assess the disease with appropriate techniques.

The choice of a particular imaging modality depends on the expertise of the doctor and the patient's condition. After recording the image of the infected organ, the disease can be evaluated by a personal check or with the help of recent image-assisted detection techniques such as segmentation, machine-learning, and deep-learning schemes.

Implementation of machine-learning (ML) and deep-learning (DL) systems to detect disease from medical images is the current procedure widely used in laboratories and hospitals to assist doctors during mass screening processes. It offers expected results on a range of disease. The choice of ML or DL depends on the following requirements: (i) knowledge, (ii) computation cost, (iii) speed of response, and (iv) accuracy. The literature also confirms that implementation of ML and DL systems considerably helped the medical community in detecting COVID-19 with higher accuracy. Further, the merit of the Artificial-Intelligence (AI) system-based detection will automatically learn during the training procedure and will help achieve better results during testing.

This book presented an overview of some selected abnormalities among humans, recording essential images of the condition, traditional and soft-computing-based image enhancement techniques, thresholding procedures, segmentation approaches, and the implementation of the DL schemes to analyze the greyscale and RGB-scale pictures collected from various datasets. This book also discussed information on various datasets and its locations. The presented methods can be implemented individually to analyze the image or it can be combined to form a hybrid system for better assessment of the abnormality with the help of medical images.

Index